Soil Fertility, Renewal

SOIL FERTILITY, RENEWAL & PRESERVATION

Bio-dynamic Farming and Gardening

by
EHRENFRIED PFEIFFER

with an introduction by
E. B. BALFOUR

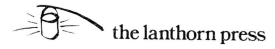 the lanthorn press

Peredur, East Grinstead,
Sussex, England.

First published by the Anthroposophic Press, New York
Revised edition published 1947 by Faber & Faber Limited
Second impression 1949
This edition published by The Lanthorn Press

Printed in Great Britain by
Penwell Ltd., Parkwood,
Callington, Cornwall

© The Lanthorn Press 1983

Cover design by Arne Klingborg

ISBN 0 906155 12 6

All rights reserved. No part of this publication may be reproduced, stored in a retrieval system, or transmitted, in any form or by any means, electronic, mechanical, photocopying, recording or otherwise, without the prior permission of the publishers.

Foreword

This valuable handbook on the Bio-Dynamic method of agriculture inaugurated by Rudolf Steiner, published in Great Britain in 1947 by Faber and Faber, has been out of print for over two decades. Widespread demands have led to the decision to print this facsimile copy of the original text.

Lady Eve Balfour, that great figure in the forefront of the international organic movement, has kindly consented to add a new and updated introduction to the one she wrote at the time of the book's original publication.

Further additions are the Notes contributed by Katherine Castelliz in Appendix A and a list of related literature in Appendix B. No attempt, however, has been made to up-date financial assessments or statistics contained in the text as these are not essential to the main substance.

We wish to thank Messrs Faber & Faber Ltd for their ready permission for this edition, and Mrs Adelheid Pfeiffer, the widow of the author, for her help and encouragement.

Michaelmas 1983   JOAN RUDEL

New Introduction

I very much welcome the republication of this book. The principles expounded therein, and the experimental and scientific data presented in support of the viability of their practice, are as valid and relevant today as they were when the first English edition was published in 1947.

Only when it comes to some of the author's recommendations for practical farming techniques does one realise that parts of the book are out of date for modern farming conditions.

It was, rightly I think, decided not to try to update by rewriting any of the text, but where any techniques described are no longer applicable, the methods whereby the modern Biodynamic farmer applies a given principle have been supplied by Miss K. Castelliz, in footnotes.

In addition to the need for such occasional updating, a few ordinary terms are sometimes used with a different meaning than that normal today. This perhaps needs a word of comment; for example, housing for *any* farm livestock Dr. Pfeiffer calls a stable, a word which, to the ordinary English reader, denotes housing for horses only. Another example is that the word "Fertilizer" is used throughout the book to cover all additions to the soil including manure and compost. This is slightly confusing at first. However, where such unfamiliar word-usage occurs, studying the text will usually reveal the correct meaning.

Some of the quoted improvements in quantity and yields following conversion of a farm to Biodynamic methods are striking and convincing, though I feel bound to mention that from time to time I have seen equal success resulting from conversion to "ordinary" organic farming, which does not normally involve the use of special preparations. However, both methods are based on wholeness and the "dynamic" concept.

Chapter 11, on Dynamic Activity is a splendid exposition of

INTRODUCTION

this concept, and includes a plea for a Dynamic Botany (p.109). This chapter alone effectively scotches the myth that the only factors influencing plants and plant growth are the soil, and what we add to it, and the weather. If the reader picks up this book just to browse, he could do worse than go straight to this chapter. I shall be surprised if it does not stimulate him to read the rest.

Chapters 12 and 13 are also very important, in my view, containing, as they do, a great deal of well documented evidence of scientific findings, and of the effect of B.D.-grown food on health, and Chapter 15, the last, on Man's Responsibility, expresses wisdom that cannot be sufficiently rammed home.

Being, myself, primarily an "ordinary" organic farmer/gardener, one or two things surprised me; using quicklime in a compost heap for example (Miss Castelliz has commented on this) and the absence of deep rooting herbs in the ley mixture recommendations, but as a founder of the original Haughley Experiment[1] I rejoiced in Dr. Pfeiffer's complete understanding and acceptance of the principles that lay behind that piece of pioneering research — an understanding that was very rare among scientists at that time. Rudolf Steiner had it, of course, and as the Biodynamic system was founded on his teaching I ought not to be surprised that Dr. Pfeiffer's outstanding perception and vision followed the same lines.

There is a good chapter of comment on forestry and one on market gardening. This is right, since both, together with farming, are part of the whole organism that constitutes an environment, the key to the management of which is balance. Balance within the parts and balance between the parts and the greater whole. Dr. Pfeiffer admits no compromise on this. On that part of the whole represented by farming he says (p.65) —

". . . Those who object to the state of balance aimed at in Biodynamic Agriculture have a complete misconception of what a farm should be. Making things grow on a piece of soil is not necessarily farming — it may only mean destroying the earth's fertility."

For the sake of readers too young to have known much about this great man's work and teaching, I think a few biographical details may be of interest.

As a young man Dr. Pfeiffer studied chemistry. In 1923 he made the Biodynamic preparations for the first time, under the

INTRODUCTION

direction of Rudolf Steiner. This was before Rudolf Steiner gave his Agricultural Course in 1924.

Pfeiffer pioneered and developed the method of diagnoses through Sensitive Crystallisation, and he dealt with questions of bio-chemistry, agriculture, nutrition and medicine. It was for his medical work that he was awarded a doctorate in the U.S.A.

He visited the U.S.A. many times in the 1930's. He went there for good in 1940, and developed a laboratory for agriculture as well as medicine in Spring Valley, New York State, where twice I had the privilege of visiting him. The spread of Biodynamic agriculture in the U.S.A. was entirely due to his work. He died in 1961.

When this book was written, peasant farming was still well-known, or at least well remembered on the European continent. Factory farming, specialisation and the tendency to make farming units as large as possible was only just beginning. The poisoning of water through spraying and nitrogenous fertilisers were hardly yet noticeable. Now, of course, a significant swing in the other direction has started, in part due to advances in biological sciences, in part to such books as "Silent Spring" and "Small is Beautiful", all of which have contributed to what is, perhaps, the strongest causitive factor of all, the ever-growing public, as well as scientific, awareness of the importance of an ecological approach to all problems connected with the environment and the living world.

It is this last fact that makes the reissue of this book so very timely. People everywhere are looking for alternative life styles, and ways of working towards a sustainable agriculture. They need in this task to be re-awakened to some of the wisdom of our ancestors and to learn how this can be adapted to meet the demands of the 21st century. It is not a question of going backwards, but of discovering where we made the wrong turn, and then going forward again in the right direction. In this vital enterprise, on which the survival of the human species may well depend, this book can be a really practical help, but let Dr. Pfeiffer himself describe its purpose. I end by quoting part of the last paragraph from his penultimate chapter.

". . . It is a familiar truism that the farmer learns more through practical demonstrations than from any amount of writing, because he is *more convinced* by what he sees than by what he reads. So this book aims to constitute a brief instruction in how to make possible this 'being able to see the

INTRODUCTION

thing in practice'. It is no textbook or directive on farming. Rather it presents the long-neglected points of view which will, when put into practice, restore ordinary farming to a healthy state which in turn will help to foster a permanently sound agriculture and a healthier human society."

E. B. BALFOUR

1 *The Living Soil & The Haughley Experiment* by E. B. Balfour, published by Faber & Faber, 1971.

Preface

Every human being should be interested in the fertility of the soil. From the earth he obtains his food. If he is a producer whose income is derived from crops and whose work is the tilling of the land, the subject of soil fertility is one of particular importance. For him the quality and flavour of the produce of farm or garden are the material bases of both his livelihood and his health.

The methods employed in agriculture in any society are important not merely to the farmer. They determine the quality and nutritive value of the foodstuffs produced and consumed and they affect, as well, the cultural and social life of men living together. "Culture", in its original significance, means work upon the earth just as it means, in its broadest sense, all that has been achieved by the human spirit. Races standing on a high cultural level have nearly always had a well-developed agriculture.

To-day agriculture is regarded merely as one aspect of the economic life. The ancient picture of farming has been altered by economic considerations. The farmer's methods of cultivation and the whole manner in which he regards his agricultural unit have changed along with the growth of industrialism and the spread of technical knowledge. The tilled field has its productive capacity and its fixed charges calculated to the last decimal. The nutritive substances in the soil have been analysed by scientific research. They are weighed and balanced against the nutritive requirements of the plants. As far as the mineral content is concerned, the most comprehensive studies have been carried out. As a result of all this, the so-called intelligent farmer, since the rise of chemical science in the middle of the last century, finds himself more and more forced to assume the role of the proprietor of a "growth factory". His means of production are his soil, his farm implements and the growth-capacity of his plants.

The economic output of agriculture is calculated like the output of a machine, despite the fact that agriculture and manufacturing are essentially different. The output of agriculture is relatively lower than that of the machine, for in agriculture a great amount of energy must be applied for a very modest return. When the production of a manufactured article ceases to pay, the production is stopped and the means of production are sold, destroyed, or put to other uses. Such a step is not possible in agriculture as a whole. In farming, when bad times wipe out the farmer's already modest return, the farm enterprise cannot be wound up and disposed of so simply. One abandoned farm in a well-cultivated region does a great deal of harm to the whole neighbourhood, for example, in the spreading of weed seeds and the change which takes place in the quality of the land. A number of such abandoned farms near one another can even lead to the devastation of a whole district and to national catastrophes. Examples of such destructiveness are to be found in ancient times in Mesopotamia and to-day in the dust bowl of North America. The kind of thinking used to calculate machine production cannot be applied in agriculture.

The truth is that in manufacturing we are dealing with something primarily inorganic. Its general calculability, as well as the calculability of its individual factors, are all easily controlled. Agriculture, on the other hand, works with living factors, with the growth, health and diseases of plants and animals. It has to do with the enlivening of the soil. All of its factors are variables. In their individual characteristics they are independent of one another, yet they unite to form a higher unity, a whole, that is to say, *an organism*.

Raw materials are received by the factory and are transformed into finished goods. Between these two poles in manufacturing—the pole of the raw materials on one side and of the finished commodity on the other—there stands the machine. The machine is not a variable factor except for deterioration. Agriculture, on the other hand, has for its one pole fertilizer and seed as raw material; it furnishes vegetables, grain, fruit, etc., as the finished product. But between the beginning and the end of agricultural production stands the *life process (biological process)*. Economic thinking could form a correct idea of what takes place in agriculture only if this life process could be taken into its calculations.

The purpose of this book is to stress the importance of this life process and to show, as the practical experiences of the author have also shown, that the farm or garden must be considered as a biological organic unit, rather than as a series of unconnected processes. When the farm or garden is operated on this basis, it will be seen

PREFACE

that *that which is biologically correct is also economically the most profitable*. The book also purposes to show how the biological unity of the farm may be achieved and developed. Its contents are based on the author's own experiences of many years in dealing with the problems of his own farms, as well as with those of hundreds of other farms and gardens in practically all countries of Europe. The author has also studied farming conditions in North America, Egypt, and Palestine.

Thus the author believes that he has been able to arrive at a comprehensive judgement of the possibilities of a method of agriculture which has become known as the *Bio-Dynamic Method*. Its originator, Rudolf Steiner, gave the basis on which this book rests. To him our profound thanks are due. We are also grateful to those who have taken the indications of Rudolf Steiner and put them into practice for the last twenty years in their respective fields as farmers, gardeners, foresters or in purely scientific research. Already there are over two thousand persons occupied in this work.

The reader will find references to certain "preparations"—numbers 500 to 508—which are an essential feature of bio-dynamic farming and which have become known to farmers by these numbers. Certain of these preparations are used in the manure and compost heaps to hasten the rotting process and to give it the proper direction. Two of the preparations are sprays for the soil and plants. The experienced bio-dynamic farmer can make these preparations himself or may obtain them from other bio-dynamic farmers. The substances from which they are made are referred to later in the book.

Although anyone wishing to start bio-dynamic farming may have these preparations with instructions for their use, the directions for making them are reserved to bona-fide bio-dynamic farmers of standing. This apparent secrecy is observed in order to prevent these preparations from being commercialized or from being made by incompetent persons. No bio-dynamic farmer (or anyone else) is permitted to make a financial profit from them.

There are now bio-dynamic farms in nearly all English-speaking countries. In some places associations of bio-dynamic farmers have been formed for discussion of problems and mutual assistance in making the preparations available.

The experiences of bio-dynamic farmers in all fields have been gathered together at the end of this book. In this way the reader will not be burdened with unnecessary quotations and references in the text.

PREFACE

Those interested in the point of view presented in the following pages may pursue the subject further by getting in touch with bio-dynamic information centres. If such cannot be located the author will be glad to put inquirers in touch with sources of further information.[1]

To-day, seven years since the first edition of this book was written and after immeasurable suffering all over the world, the basic idea of this book still holds true, in fact more than ever. What we can add is more experience, more material illustrating the truth about organic life on this one earth, and a greater conviction of the necessity for the preservation and maintenance of the soil's fertility.

This war has made a dreadful incision into the fertility of the earth by excessive production for our needs, by deforestation, by a multitude of devastating actions, by lack of essential labour on farms.

The recovery and increase of the organic resources of the land demands our attention as one of the most important and peaceful tasks for the sake of future generations.

EHRENFRIED PFEIFFER.

Meadowbrook Farm,
Chester, N.Y., U.S.A.

---

[1] In Great Britain, information can be obtained through the Bio-Dynamic Agricultural Association, Woodman Lane, Clent, Stourbridge, West Midlands.

CONTENTS

FOREWORD	page i
NEW INTRODUCTION BY E. B. BALFOUR	ii
PREFACE	5
INTRODUCTION BY E. B. BALFOUR	11
1. THE FARMER OF YESTERDAY AND TO-DAY	15
2. THE WORLD SITUATION OF AGRICULTURE	19
3. PROSPERITY, SECURITY, AND THE FUTURE	26
4. THE FARM IN ITS WIDER CONNECTIONS	31
5. THE SOIL, A LIVING ORGANISM; THE "LOAD LIMIT" IN AGRICULTURE	38
6. THE TREATMENT OF MANURE AND COMPOST	49
7. MAINTENANCE OF THE LIVING CONDITION OF THE SOIL BY CULTIVATION AND ORGANIC FERTILIZING	62
8. HOW TO CONVERT AN ORDINARY FARM INTO A BIO-DYNAMIC FARM	69
9. COMMENTS ON FORESTRY	89
10. COMMENTS ON MARKET GARDENING	96
11. THE DYNAMIC ACTIVITY OF PLANT LIFE—SOME UNACCOUNTED CHARACTERISTICS	107
12. SCIENTIFIC TESTS	130
13. FERTILIZING; ITS EFFECTS ON HEALTH	148
14. PRACTICAL RESULTS OF THE BIO-DYNAMIC METHOD	164
15. MAN'S RESPONSIBILITY	182
BIBLIOGRAPHY	187
INDEX	192
APPENDIX A	197
APPENDIX B	199

LIST OF ILLUSTRATIONS

PLATES

1. A BIO-DYNAMIC MANURE HEAP	*facing page* 54
2. CROSS SECTION OF A BIO-DYNAMIC MANURE HEAP THREE TO FOUR MONTHS OLD	54
3. A MODEL COMPOST YARD	55
4. A HEALTHY GREENHOUSE TOMATO PLANT	102
5. ROOTS IN AN EXPERIMENT SHOWING THE INFLUENCE OF THE CLAY, MANURE, AND SAND MIXTURE	103
6. LUPIN EXPERIMENTS	132
7. LUPIN EXPERIMENTS [CONTINUED]	132
8. LUPIN EXPERIMENTS [CONTINUED]	133
9. NITROGEN BACTERIA NODULES. LEFT: USUAL TREATMENT; RIGHT: BIO-DYNAMIC TREATMENT	133
10. A SPADEFUL OF BIO-DYNAMIC COMPOST 3 MONTHS OLD. NOTE THE GREAT QUANTITY OF EARTHWORMS	134
11. ROOTS SHOW PREFERENCE FOR BIO-DYNAMIC COMPOST. LEFT SIDE OF ROOTS GROWING IN SOIL WITH ORDINARY COMPOST	135
12. (A) RADISHES TREATED WITH OAK BARK, PREPARATION 505, DANDELION, PREPARATION 506	138
(B) CONTROL RADISH, RADISHES TREATED WITH STINGING NETTLE, AND WITH PREPARATION 504	139

FIGURES

1. MANURE HEAP AND COVERING	*page* 51
2. METHOD OF STACKING MANURE IN THE FARMYARD	53
3. DRAINAGE SCHEME IN A MANURE HEAP	56
4. RIGHT AND WRONG WAY OF GROWING PLANTS ON A COMPOST HEAP	57
5. CONSTRUCTION OF A COMPOST HEAP	59
6. "MODUS OPERANDI" OF EXPERIMENT	135

Introduction by E. B. Balfour

"Knowledge of the underlying principle of human ecology", declare Jacks and Whyte in their now famous book, *The Rape of the Earth*, "is one of the most urgent needs of mankind." They define human ecology as "the art of living together with animals, insects, and plants". They should have added "and with each other". With that fuller, and more accurate definition, it is surely no exaggeration to say that knowledge of that art is to-day the *most* urgent need of mankind.

The present book is an important contribution to the study of ecology. It was originally published in America under the title of *Bio-dynamic Farming and Gardening* and I welcome this (revised) English edition as long overdue.

The advance in science over the last hundred years has been astonishing and wonderful, but the technique of the scientist, which is study through fragmentation, while having enormously increased the sum total of human knowledge, largely fails when applied to biology—the science of living organisms. You can discover of what an egg is composed by taking it to pieces and analysing it, but you cannot thereby cultivate a chicken, nor determine how the innumerable factors governing its environment may affect its health and growth, for a living organism is not just the sum of its parts, it is an inter-related functioning whole.

In agricultural research the chemist's investigations are usually undertaken within the confines of his laboratory, or at most restricted to areas no larger than the small field plot. One such investigator has even been known to state that the only research results of any value are those which can be demonstrated in a test-tube. The biologist on the other hand, recognizes that laboratory results are not necessarily valid under field conditions and he studies his specimens in a wider and more natural environment. The ecologist still further extends his field of investigation. He

INTRODUCTION

regards a whole farm as his smallest unit, and his research programme has thus to be planned on a very long-term basis. This is as it should be, yet after reading this book I was left wondering whether even this is not too fragmentary an outlook, and it was vividly brought home to me why it is that ecology—the science of the study of the Whole—is not only the newest, but also the most difficult of all the sciences, for the further the advances we make in it, the wider does our horizon become, until, having started with the idea of a single living entity as our unit, we begin to catch glimpses of a vision—already a reality to the author—that the "Whole" which we have set out to study is nothing smaller or less formidable than the entire universe!

Dr. Pfeiffer builds up his case logically and presents his evidence convincingly with a wealth of supporting, and often fascinating, statistical data. To what extent the reader will be prepared to accept all his deductions will depend upon the distance that he has already travelled along the ecological road, but the book cannot fail to provide any serious reader with deep food for thought. Lovers of the soil everywhere will intuitively recognize much that they have always known to be true, even if they consciously learn of it here for the first time, and no research worker worthy of the name will fail to perceive the importance of Dr. Pfeiffer's observations and interpretations.

In our world, which is held together by opposing stresses, it is seldom that evil does not bring some good, or good some evil. The evil which science has brought to agriculture has resulted from its setting out to be, not an interpreter of ancient wisdom, but a substitute for it. As a substitute it has been a lamentable failure. It set out to double our crops; over wide tracts of country crop yields are declining. It set out to destroy the pests and diseases of our crops and livestock; these are on the increase. When science was first applied to agriculture the traditional wisdom of the husbandman, built up through the ages by careful observation of natural phenomena, was condemned by science as superstitious nonsense and discarded. So the farmer died and his wisdom with him, and the agriculturalist took his place. This unfortunate, deprived of his faith in, and knowledge of, ancient wisdom, has to rely on science alone, and science has let him down. It is time that we retraced our steps and tried again. By this I do not mean that we should go back to blind acceptance of the so-called superstition of our ancestors, nor do I mean that we should scrap the scientific knowledge we have gained; far from it, but we should examine again the

INTRODUCTION

beliefs of our forebears and study the observations on which they were based, and we should use our new scientific knowledge to interpret those observations and to sift those beliefs.

This is the approach which Dr. Pfeiffer makes to research and to the problem of soil fertility, which is the problem of life itself.

The truly scientific mind is an open mind. Not accepting blindly, but equally not rejecting blindly. Above all it is a humble mind, recognizing how much more numerous are the things we don't know, than the things we do know, and that even the things we think we know are but half truths.

It is the unscientific mind—possessed, alas, by too many self-styled scientists!—that instantly dismisses as superstition, magic, or even as non-existent, happenings brought about through the operation of some natural law which we do not yet understand.

Those who think they know everything can learn nothing. Those who know they know little, will learn much. Scientific "truths" are always being modified as our knowledge increases. It is well to remember that, and also the fact, as L. le Mesurier puts it, that "wisdom is always more than knowledge, but never contradictory of it".

E. B. Balfour.

CHAPTER ONE

The Farmer of Yesterday and To-day

True husbandry is on the point of disappearing. Looking back over the last few centuries we can follow its decline step by step. Old traditional customs are no longer understood and practised. The whole attitude to farming has changed. In times gone by, after a week of hard work, the farmer used to walk through his fields on Sunday. His heavy, swinging gait, acquired by walking behind the plough, betokened his train of thought. It expressed a deep penetration into the processes of nature. On these Sunday walks, the creative work of the week was thought over in terms almost to be compared with the biblical epic of creation.

The farmer was often accompanied by his son, whom he initiated into the mysteries of nature. He described to him in simple words the manner of tilling the earth, the art of sowing. These experiences had been handed down from his forefathers from time immemorial. Outer rules did not yet exist; experience alone was his guide. From observation and tradition the farmers were able to use the course of nature as an almanac with signs and symbols. The budding of this or that bush indicated the time for the preparation of the seed furrow. Wild growing plants became guides for the right moment to do one thing or another. This instinctive certainty of the old-time farmer prompted him to take the necessary measures at the right moment by observing nature's course. This instinct has been lost. An uncertainty has arisen and now the successful neighbour is often the only guide for the farmer's work. "I see the farmer on the hill is out to-day with his ploughs," the farmer says. Thus one always looks to someone else for farming wisdom.

In older times it was the custom to rotate crops, with intervals between, during which the land was allowed to lie fallow for the sake of rejuvenating the soil. During his Sunday walk the farmer would say: "When this field lies fallow for the third time we shall

SOIL FERTILITY, RENEWAL AND PRESERVATION

have the dowry ready for our daughter." "When this cow calves for the last time then our son will be ready to leave school." Diligence and orderliness ensured a safe future. A certain feeling of comfort and satisfaction was spread over the entire field of farming activities. This atmosphere has often been described in works of poetry.

The farmer of that older time had a fine feeling for meteorological conditions. He felt in his bones—without the aches of rheumatism—every change of weather and was able to arrange his work accordingly. All these semiconscious treasures of knowledge of past times are to be found to-day only among so-called old "eccentrics".

Scientific agriculture has decidedly altered the ways of ancient husbandry. It has told the farmer to abandon his old superstitions; that he can obtain better crops by turning his fields into a growth factory. The economic development of the twentieth century has transformed the farmer into the agriculturist who has to "calculate" costs and output. Considerations of "profitableness", born of the decline of the old agricultural tradition, have become his "daily bread". Thoughts like the following now fill the farmer's mind: "How much will this work cost?" "What will be the yield of this field?" "Will it pay to have this field hoed again?" Many similar problems occupy his thoughts.

Although the farmer was told that by the use of scientific methods in agriculture he could double the yield of his farm, the fact remains that after years of scientific help present-day farmers are discovering that their "double" yields to-day are no better than the single yields of previous times.[1]

The work of the farmer is not dependable. In olden days he knew that if he sowed the seed at a given time, according to the moon phases, he would have a dependable crop. Now such a procedure is called nonsensical. Unfortunately modern science has developed no exact rules to take the place of the old farming wisdom indicating the time and manner of sowing. The farmer has constantly to complain of the uncertainty of his crops. He can no longer predict with any certainty the course of their growth. Earth and plants have become more sensitive and erratic. Older farmers

---

[1] Compare 'Die chemische Industrie (*Chemical Industry*) No. 11, 1933, pp. 166. Average use of nitrogen as fertilizer in 1,000 tons 1912–14: 170; 1927–9: 408; 1933–5: 387. Average yield of 100 pounds per acre (for summer barley) 1912–14: 21.3; 1927–9: 20.0; 1933–5: 19.8. The figures given here are somewhat similar for oats.

THE FARMER OF YESTERDAY AND TO-DAY

tell how things have changed. And, because of economic pressure to-day, farmers have no time for making experiments.

"The calculation of 'profitableness' ", said the German national economist, Werner Sombarth, "is an invention of the devil by which he fools human beings. It has destroyed a colourful world and transformed it into the grey and gloomy monotony of money valuation."

Thus the "farmer" has become an "agriculturist". Development, under the pressure of "profitableness", has forced him to resort to the use of machinery to replace the more expensive human labour. For a certain length of time relief was apparently obtained by this method, particularly through the use of harvesting machinery. But in the case of machinery devised for the tilling of the soil this "relief" has only been partial. Here the value of intensive manual labour has proved itself indispensable; it cannot be replaced by machines. Machines accomplish their results in a shorter time, it is true, but also in a more superficial way. The fine humus structure of the soil, which once existed, cannot be produced or preserved by machines. Those who treat the earth with a "pulverizer", for example, destroy the real creators of natural humus, *the earthworms*. The machine age has brought about great changes in the work of men whose business it is to till the soil. It has created the "growth mechanic" type of farmer. This age has also seen the ever increasing use of mineral fertilizers. At the same time the yield of crops which ascended at first has ceased to climb in the same way with the continued use of these fertilizing methods. The situation to-day shows that while, in comparison with the time before the 1914–18 war, three times as much nitrogen is being used the average yield per acre has not increased. In some areas it has actually decreased. Yet the idea that there may be an uneconomical principle underlying the fertilizing methods now in use is often considered heretical. Another problem of the times is the increase in the phenomena of degeneration—plant diseases and insect pests. A description of these phenomena is not necessary, for they are a part of the farmer's daily experience.

We need only ask ourselves how many of our present working hours have to-day to be devoted to spraying against pests and to seed baths, and how many hundredweight of the yield are destroyed by pests once the harvest is gathered. Think of the role of the corn billbug and granary weevil. Experts tell us that one half of all that starts to grow is lost. The final result of all this is that crop costs are higher to-day without any higher crop yields than was the case thirty or forty years ago.

SOIL FERTILITY, RENEWAL AND PRESERVATION

Recognition of these facts causes many questions to arise in the mind of the farmer. If he were fully alive to the seriousness of his situation, he might say to himself when he walks over his fields: "What has happened? How can this unwholesome development be stopped? I am witnessing the sickening of the whole earth. This earth sickness cannot be stopped by isolated measures taken on a single farm."

The ways and means for the regeneration of the farm can be found only in a comprehensive view of the earth as an *organism*, as a living *entity*. It is clear that the methods of so-called "scientific agriculture" with its striving for ever increasing yields and its continued use of ever more mechanical help has led to an impasse.

CHAPTER TWO

The World Situation of Agriculture

It is not necessary to go back to classical antiquity to study the transformation of the most fruitful land into desert because of false or one-sided utilization. The erosion and dust storms of the American Middle West serve as a dramatic example of the present day. Adjoining the fruitful black soils of the Missouri-Mississippi Basin, a broad plain extends perhaps 500 or 600 miles westward. Bordering this plain is cultivated land, then comes what has been for centuries prairie. On the west, this region is bounded by the Rocky Mountains. Once this prairie was the home of the wild buffalo; later it became a cattle pasture region, covered with the heavy sods of prairie grass.

Because of an unbalanced overgrazing and a greedy tillage of the soil, with no care given to protect it, this sod was gradually loosened and thus the protective soil covering was destroyed. Drought and wind then accomplished their task, and so it came about that this top soil began to "wander". In the cultivated sections, intensive and soil-consuming grain culture, without beneficial crop rotations and harrowing, without rolling at the right time, and with many other errors in tillage, has all contributed to the loss of the organic substance of the soil once so rich in humus. The top soil has disintegrated and the capillarity of the soil has disappeared, thus further advancing the erosion.

Experiments at the Missouri State University, in Columbia, Missouri, have shown that the nitrogen and the humus content of the soil, when contrasted with the original prairie, have fallen about 35 per cent. In about thirty years, the fertile condition of the soil has changed to such a degree that, according to Professor Jenny,[1] the only thing which has increased during this period is soil acidity. To-day it would cost more than the land is worth were

[1] *Soil Fertility Losses under Missouri Conditions*, Columbia, Mo., 1933; Bull. 324, Agr. Experiment Station.

anyone to attempt to bring it back to its original state with artificial aids; that is to say, practically considered, such restoration is no longer possible. The dust storms which have been prevalent in recent years are completing the process of this retrogression of fertility. Many other states are seriously affected and thus *a third of the cultivated area of the United States of America is on the way to becoming useless.*

In addition to these external natural catastrophes, economic difficulties have overtaken agriculture. The American farmer is faced with the dilemma either of employing as few helpers as possible or of abandoning the farm altogether. Human labour, because it is too expensive, has had to be replaced by machinery. But the mechanical working of the soil can never conserve humus as effectually as human manual tillage. The soil is now hand-tilled only as much as is absolutely necessary. The biologically beneficial form of the balanced, diversified farm—with heavy legume plantings, meadows, green manuring—was given up in favour of a one-sided cultivation. The result of this course—a policy lacking in foresight, helped along over a few years by poor weather conditions—has been that one of the great grain-producing areas of the world, the United States, has imported cattle feed, and that in 1936 this wheat-exporting country actually imported wheat.

The conditions are similar in Canada where within recent years a 50 per cent harvest was reported. And if such conditions continue it is within the bounds of possibility that two of the most important grain-producing countries of the world will soon be living "from hand to mouth".

If we look now at European conditions before the war, we see that there, too, we can hardly talk about bumper wheat crops. Sir Merrick Burell, chairman of the Standing Committee of the Council of Agriculture for England, presented the following resolution in July 1936: "The Council of Agriculture for England is seriously concerned over the position of the national food supply, which is obviously at present one of the weakest links in the chain of National Defence. Both the fertility of the soil and the means for increased production of foodstuffs are to-day less than they were in 1914 and subsequent years, when the shortage of food supplies placed the country in a most perilous position. The Council, therefore, desires to urge on the government and, through it, on the country the great national importance of this question. It suggests that those responsible for defence should be requested to give it their immediate attention so that remedial measures may be at once set on foot." The resolution also warned the Cabinet that the *lowered reserves* of fertility in the *greater part* of the soils of the country and the fewer men employed to-day made a rapid expansion of the production of foodstuffs absolutely impossible. Only a carefully thought out, *long-term* agricultural policy, embracing all sections of the industry, would be likely to prove adequate.[1] The War Departments in various countries often have a broader view of the agricultural realities than have the agricultural experts.

In the middle of the last century there was a flourishing agriculture in England, while to-day there are wide areas covered with heather and ferns, and the acidity of the soil is constantly increasing. If we consider this carefully, we find that here, also, the conditions are in principle no different from those in North America. We can observe how, under the influence of a moist climate, the pastures and meadows are taking on an increasingly mossy character. Social and economic conditions are also doing their part towards promoting a one-sided cultivation of the soil. It is beginning to dawn on the persons concerned that a one-sided agriculture can produce only temporary results.

There was one picture in England that impressed the author especially. He visited a farm with a heavy clay soil. The analysis of the soil was discussed, and a complete lack of calcium was noted. According to the rules of the mineralizing theory, these fields should have been immediately spread with lime. But the peculiarity of the situation was that the surface of the soil—and such cases are not rare—was from one and one half to two feet deep, and under it there was a deposit of practically pure limestone. The top had separated itself completely from the subsoil, carrying on its individual existence like a thick rind. There were no longer any biological connections between the life of the plant roots, the soil bacteria and the earthworms. One-sided measures of cultivation had brought about this isolation. The earthworms usually go down as far as six feet and more into the ground, and carry lime up from the lower levels provided they receive the proper stimulus for the formation of humus. Also the roots of the legumes, if these were planted, would penetrate deeply into the earth and through their capacities for making the soil elements available—capacities which are forty to sixty times greater than those of the grains—would bring about and stimulate an interchange of elements. A properly timed aeration would also help such a process of healing, provided

[1] *Journal of the Ministry of Agriculture*, vol. xliii, No. 4, July 1936, p. 372; Report from the 'Council of Agriculture for England.'

SOIL FERTILITY, RENEWAL AND PRESERVATION

the first steps toward the formation of humus substances were taken.

In a similar way, processes causing a continuous reduction of soil fertility are taking place on the whole European continent.

If we turn to the Far East, we see that whereas the "manless farm" represents the typical picture of the Occident the opposite is the case in the Orient, in the great river valleys of North-East China. Overpopulated farm areas are typical there. While the United States has 41 inhabitants to the square mile, Switzerland has about 225 to the same area, Germany 343, England 660. Censuses of the heavily populated regions of China and Japan show from 1750 to 2000 inhabitants to the square mile.

In the province of Shantung, a family of twelve plus a donkey, a cow and two pigs is normal for a farm of two and a half acres. In Central Europe a peasant farm of forty acres, about sixteen times the size of the Chinese farm just given, can barely support a family. In the thickly settled Asiatic regions 240 people—besides their domestic animals—live on the same amount of ground and on what it produces. Obviously the extraordinarily low requirements of the oriental peasant and agricultural labourer make this possible. Yet from the biological point of view the productivity of a soil which for thousands of years has been able to give complete nourishment to its inhabitants is an amazing phenomenon.

The intensive Chinese cultivation of the soil rests on a humus and compost economy carried on with almost religious zeal. Everything that can be turned into soil is composted: plants, all sorts of refuse, the muck of streams, plain dirt, are all set up in layers, kept watered, and in a short time turned into humus earth. All work in China is manual labour. This conserves the soil and permits an inner aeration and mixing. Mixed cultures of as many as six different kinds of plants in various states of growth utilize the mutual action of plant groups upon one another. To enlarge the surface area, everything is planted in between and upon ridges. Mineral fertilizing is still unknown there—fortunately for the Chinese. Scientists who have visited such regions and studied them from the background of their technical knowledge say that a crop failure in this soil is a rare phenomenon.

Here, then, it has been possible to keep a land in its original state of fertility by the use of the oldest cultural methods of humanity—*humus conservation* and *manual labour*. And yet the unnatural overpopulation shows that here, too, the biological balance is disturbed. A visitor to the great river-valley plains can observe on adjacent hills and mountain chains poorly grassed or steppe-like impoverished land, rapidly approaching the characteristics of a desert. The overpopulated fruitful regions are immediately adjacent to the extreme opposite—unfertile desert lands.

We know that it was lack of food, caused by the overcrowded conditions of their country, which really drove the Chinese to cherish ancient, tested methods of cultivation with religious reverence. The value of the ground to the individual who inherits it can be seen in many a law and regulation. The land is the property of the State, formerly of the Emperor. Whoever *cultivates the land well can call it his own as long as he cultivates it.* If he ceases to do this, then another person, or the State, has the right to take over the field in question. If one of the numerous floods carries off a piece of land from the river bank, the peasant has the right to "follow" it. Should he be able to locate the spot where the soil and mud have been deposited, he can settle there. If this misfortune happens in an already settled region, then the earlier owners, and the owners who follow the earth carried down by the flood, have to divide their small piece of land.

When the soil of the field begins to show signs of exhaustion, the whole family co-operates in carrying the top soil in baskets and carts into the farmyard. There the earth is carefully mixed with manure and plant refuse and composted. While there, it is turned over many times. Meanwhile the lower soil of the field is reinvigorated by a planting of legumes. After a time the regenerated earth is again returned to the field and the cycle of thousands of years begins again.

If we can speak in the West of "natural" catastrophes which threaten the regions of human habitation, we must in a similar way point to the East where other human catastrophes of the most tremendous magnitude are in the process of preparation. The overpopulated land in China has already reached the extreme limit of its productive capacity for meeting even the minimum existence standards of the people living on it. Next to this land lie the poor hilly regions, stripped bare of their trees. The process of deforestation in China had already begun on great stretches of land as early as 1000 B.C., and was practically completed by A.D. 1388. The only exceptions have been some southerly and south-westerly regions which in part have a climate approaching the subtropical. The general conditions of the water supply of a country are to a large extent disturbed as soon as the balancing effect of wooded areas is lacking, and without these wooded areas the extremes of

SOIL FERTILITY, RENEWAL AND PRESERVATION

climate draw still farther apart. Sudden torrents of rain and floods alternate violently with periods of drought, heat and extreme cold. The balancing, water-retaining, cooling and warming regulation provided by the woods is lacking.[1] And the lack of wood in China brings another problem with it, the impossibility of getting fuel for cooking and for warmth.

One need have no prophetic gifts to realize that in those regions a human catastrophe is in preparation. This will have an active effect on the inhabitants of Europe as well. Dr. Steiner once suggested that the student of human affairs would do well to investigate the influence of food as the driving, unrest-producing motive power behind the migrations of peoples. Races which had remained for long periods of time on the same soil have felt them-

---

[1] The following is quoted from the exceptionally fine description by Professor G. Wegener in an article on China in the *Handbuch der Geographischen Wissenschaft*, 1936:

"On the slopes of the hills we see either a dusty, poor stand of plants of secondary quality, or the complete and fearful bareness and emptiness of stony ground. The higher portions are quite badly furrowed and torn by the rain, while the lower are always newly covered with streams of mud which destroy the fruitful soil at the foot of the hills. The levels and valley bottoms are cultivated as gardens right to their extreme edges. The great lack of fuel drives the inhabitants to a ruthless interference with the natural reforestation, even on the uncultivated lands. Richthofen recounts with a sort of rage how in North China there is widespread use of a kind of clawlike utensil by means of which even the most modest little plant root is torn out of the soil In some sections of the country grass roots, dung, etc. already to-day have to serve as the only fuel. In South China, although this was settled much later by the Chinese and is still to-day much more thinly populated, the exhaustion of the woods is also already very far advanced and goes forward at a gigantic pace; and here, too, with all its devastating results. This is especially true in the area of thickly populated sections, as well as in the region on the Sikiang Delta. The Chinese culture, which in its conservatism is still to-day very strongly a culture of wood, needs this wood in large amounts—and so the consumption of the forests goes ceaselessly on. It thus becomes clear that even for the inexplicably cheerful Chinese peasant the minimum standards of existence are being rapidly left behind, and that precisely from these provinces comes the main stream of the tremendous Chinese emigration which Manchuria has in recent years absorbed. The denuded hills and mountainous regions, about 50 per cent of the total area, remain practically unutilized.

"From here, as well as from the similarly oversettled regions of China," says Professor Wegener, "there comes forth a type of man who, because of his low standards of living and the toughness of his body and nerves for the performance of his traditional labour, eliminates every competitor from the field both at home and in the lands in which he settles. Herein lies perhaps the greatest 'Yellow Peril', when the spreading of the Chinese over the earth becomes more extensive than to-day!"

selves "becoming restless" in the development of their common life because of their one-sided nourishment, and have sought for a balance.

What, in short, is the situation of Europe which occupies a middle place between the two extremes of the East and West? In Europe there is a temperate climate, and a healthy division between woods and fields, lakes and plains, hills and moorlands. There was still in the last century a peasantry firmly based on the soil, one which tilled the earth out of a traditional wisdom. But engineering methods and science, together with economic difficulties, have already taken hold of the situation, so that the originally healthy structure of the soil is beginning to fail. The place of the peasant is being largely taken by the traditionless agriculturist, and the "tiller-mechanic" who appears no longer to have any relationship to the problem of the soil which is being attacked by a world-wide sickness.

Whoever is candid and honest with regard to the facts of modern agriculture realizes that this situation exists, although he may not like to face it. The peasant knows it by his feelings. It is known by the research scientist on whose desk reports pile up concerning the unsolved problems of soil fertility, the fighting of insect pests, etc. A leading agriculturist said not long before the war: *"We have been exerting ourselves for the last two decades to halt this process but our efforts have been fruitless. What is the solution? Where is the answer to be found? The disease caused by lowering the natural soil fertility is a world sickness."*

CHAPTER THREE

Prosperity, Security, and the Future

"Prosperity" was the rallying cry of the mechanical-technical development of the past century up to the time of the first World War. It became "Security" during the inter-war years and this has lasted to the present day. "Self-maintenance" may be the motto of the coming decades. The Western countries, whose most progressive representatives can no longer believe in a prosperity based upon a constantly enhanced production, seem to think that a timely transformation to such a basis of individual self-maintenance may still save their economic existence. War economy has destroyed the products of increased activities and will once more prove itself an illusion as a means to economic betterment of any nation. Only with utter slowness does mankind take to the laborious way of looking for spiritual guidance in solving its urgent problems.

The impulse of "self-maintenance" can be met, strange though it may seem, in the thoughts of the most varying and frequently contrasting economic systems. All countries have the idea of self-sufficiency in common, at least in the sphere of food, and it is based on the fundamental truth that if one is unable to support oneself one cannot live and work for others. This is demonstrated in a simple way by developments in agriculture.

An interesting analysis of American conditions by O. E. Baker reaches the conclusion that a self-supporting agriculture offers the only possibility of maintaining the fertility of the soil. Ethnically, this provides new generations for industry and cities; economically, it acts as a stabilizer because properly managed farms are able to exist and pay their taxes in times of economic depression and unemployment. In other words, the agricultural soil capital can then at least defend itself and earn a small yield of interest. Of course, the profits and standard of living are so modest that anyone dependent upon modern economic habits of life can hardly feel attracted to such farming. Nevertheless, it means some progress, in comparison with the prosperity-security age, to be able in this way to earn a modest living.

The return to the self-maintenance principle seems essential for two reasons, namely unemployment and increasing satiety with industrial and city life on the one hand, and, on the other, the increasing destruction of the soil's fertility, which can only be prevented through the creation of small and medium-sized mixed farms. Whereas even ten years ago authoritative representatives of agriculture stated that the farmer first of all must make a profit, the Department of Agriculture in the United States has since fundamentally changed its point of view. Secretary Wallace announced in December 1939 (according to the *New York Herald-Tribune*, 26 December): "The most important change in the programme of the Farm Security Administration foresees that no credits will be granted if the farmer asking for credit intends to plant only one cash crop. In this way we hope to popularize the practice of crop rotations and mixed farming in monoculture districts, particularly in the Southern cotton districts. Farms applying soil-protective methods will obtain larger credits."

Western culture has thus reached a turning point. Ever increasing production and technical improvements no longer seem possible. Indeed, it is a question of reduction in order to re-establish proper balance. In biology we recognize "division by reduction"; for example, when the eggs mature and new cells are generated, four of these arise to begin with, of which three die, thus leaving sufficient growth substance for the surviving cell.

Growth thus gives rise to new growth, capable of life, after having eliminated everything that is superfluous. We must therefore reckon with the will to live, and with renunciation. What Nature first supplied in abundant measure is taken back again, in order to make room for new things. The forces of expansion and contracttion will be observed in ever rhythmic change. If we wish to make true progress in economic life, we must recognize such forces. The Westerner is here handicapped by mistaken ideals. He is afraid of losing freedom of trade by passing over to an economy based on forms of organic growth; he does not realize that at the present time freedom of trade exists in name only.

The agricultural forces remaining after a "division by reduction" should be led into suitable channels in order to provide for liberated human beings. In the social organism it is not only our duty, but also our right, to find work.

SOIL FERTILITY, RENEWAL AND PRESERVATION

Production must be brought into line with possibilities of consumption; for it is overproduction, with its superfluous needs stimulated artificially through advertising, that has led to the present conditions. The great task of the immediate future will be that of limiting production to actual requirements and at the same time guaranteeing every individual human being an existence worthy of human dignity. In other words, the economic system must be kept going without revolutions, unemployment or other vicissitudes. Many people still shrink from this task; they prefer to leave these problems alone, getting into wars and misery, hoping that things will somehow straighten themselves out.[1]

A great moral uplifting is needed, in every individual and in every nation, in order that everything superfluous be renounced and only what is necessary be demanded. It is obvious that this cannot take place without difficulties. A period of great economic change will be inevitable. To-day we can still decide freely whether evolution or revolution shall mould the future. The forces that

---

[1] In 1939, two months before the outbreak of World War II, the writer explained to the agricultural Director of the French Ministry of Agriculture, M. B., the ideas and experiences of bio-dynamic farming and gardening, along the lines of the principles set forth here. The necessities of maintaining the fertility of the soil and of adopting a far-sighted agricultural programme were particularly emphasized. M. B. replied: "You speak as if you wished to bring into being a dynasty of peasants; but what are we to do with those who demand to make a profit and for whose profits we shall be responsible?" Whereupon the author answered: "The fertility of the soil and sound reserves for the existence of mankind must suffer, if you do not adopt a bio-dynamic agricultural programme based upon long periods of development. The future of the French nation may really depend upon just this dynasty of peasants coming into being. If you bear in mind only the immediate profits, you will be living on the capital of the earth's natural fertility and the nation will be shortlived. The Minister of Agriculture is responsible whether immediate profits during a relatively short period of time shall be allowed to deceive the people or whether measures can be taken that will ensure the life of the nation for centuries to come."

After the French nation had passed through unspeakable catastrophes and suffering, we read the following proclamation by the French Government, dated Vichy, 3 July 1940: "MEASURES OF THE FRENCH MINISTRY OF RECONSTRUCTION: 1. France is above all a nation of peasants and craftsmen. These callings have been neglected too long and must be brought to life again. 2. A sound balance between rural and industrial activities must and shall be re-established. 3. All workmen who are not specialists, and who have been absorbed by the war industry from rural districts, must again return to the land. 4. A general policy of rural repopulation must be adopted. The earth of France can occupy and nourish far more people than has been the case during the last years." To-day (1945) this holds true more than ever before.

were fettered by war will be set free again. Many things will then cease to be produced and many workers will be idle. There will be no golden age and it is wise to begin paving the way for the direction of those forces that will be set free. The self-maintenance principle will then be a necessary expedient, until balance has been established between economic life and the social organism.

Experiences and disappointments connected with the return of the industrial and city populations to the country have given rise to much criticism. In many cases the negative results were due to lack of agricultural experience. The greatest obstacle was, however, of a psychological nature, namely the unaccustomed simplicity of country life in comparison with city life. As we have said, the financial prospects of a self-supporting farmer are not very tempting. He has his home and produces his own food, but beyond that he does not earn very much. Unemployed workmen very often prefer cash Government doles, because the farmer's bare possibilities of income may not be any greater and he has to work very hard. Some other temptation than the financial would therefore be needed. A description of possible future hardships does not induce people to change their way of life. And the change to a self-supporting agricultural life must be preceded by corresponding training and education, for no one can become a farmer or gardener merely by picking up a spade or putting on heavy boots. Another incentive will therefore be needed in order to carry through a permanent agricultural colonization programme, thus creating a balance between urban and rural life. What may that incentive be?

The born farmer, with his innate love for the soil and his especial psychical forces connected therewith and also with his professional and traditional background, is able to follow his calling in a way that townsmen returning to the country, or partly self-maintenance seeking suburban dwellers, can mostly experience only after very long practice. The first essential thing is to awaken in them a feeling for the forces of growth, for the eternally creative forces of Nature. The next step is to awaken in them a sense of responsibility towards these forces of growth, towards the health of the soil, of plants, of animals and of men, and also an inner sense of satisfaction in progressing towards this goal. This in turn becomes a compensation for the modesty of the livelihood earned. Those who cannot develop these ethical qualities will never become good farmers or colonizers, and they will hardly ever become constructive members of the social organism. The main reason for the

SOIL FERTILITY, RENEWAL AND PRESERVATION

failure of so many social experiments, and more particularly of resettlement projects, has probably been the lack of these qualities in the majority of the participants.

On the plains of Northern Italy I once visited two peasants whose fields were in an exemplary condition and whose cattle, more than one hundred milking cows and many breeding bulls, had a high value. Both men were over sixty and could neither read nor write. Then I realized that farming requires something that cannot be learned at school or through reading or degrees, but is a profession which is also a world-conception, a service to Nature, a true "calling".

If to-day economic and ethnical reasons call for an increase of the rural population (the smallholder striving for self-maintenance belongs to it, as he lives on a biological unit of the cultivated earth) this goal can only be reached if moral factors are given due consideration. The spiritual life thus comes to the fore. The industrialized, technological, scientific life of the past fifty years could never carry through such projects, as it proved fundamentally hostile to rural life. Consequently the first inevitable requirement is a change in spiritual direction.

CHAPTER FOUR

The Farm in Its Wider Connections

The farmer has not only to deal with *his* soil and with *his* seed; he is connected with an encompassing life process in his wider surroundings. One simple fact may be mentioned in this connection. Although the farmer's land is spatially limited, the plants on it draw only from 2 to 5 per cent of their nourishment from the mineral substances in the actual soil itself. The remainder—water and carbonic acid—originates from the air. But through its relation to the atmosphere, with all its currents and movements, the plant world is connected with the earth as a whole, and therefore its sustenance may come from beyond the boundaries of continents.

As a result of the movements of the air, due to wind and to high and low pressure areas, it may happen in Europe, for example, that the air comes at one time from the middle of the Atlantic Ocean, at another from Greenland, or again, from North-East Russia. A cold wave, an inrush of cold air from Greenland, has quite a different character, for instance, from one coming from North-East Russia, or in the United States from Canada. The former brings a moist cold, the latter a biting cold, but dry air. Furthermore, numerous observations have shown that the whole physiological state of an organism is dependent on these atmospheric influences. In the human body, blood pressure, propensities toward every possible sort of organic disturbance, attacks of influenza, etc., parallel the variations in atmospheric conditions. But these influences, although they are little observed, also play a role in the plant kingdom. Thus, for example, mildew may appear on spinach "overnight", brought by changes in weather conditions. The aforementioned moist air from the Atlantic, which has a tendency towards forming mist and fog, has a fostering influence on the prevalence and spread of those fungus diseases of which this mildew is an example. With such a change in the weather, it suddenly appears.

SOIL FERTILITY, RENEWAL AND PRESERVATION

Therefore in relation to climatic conditions we see agriculture linked up with a larger earth sphere. This applies to the woods, fields, prairie-land, bodies of water, moorland. All together they form in a larger sense a self-enclosed organism whose parts (woodland, field, etc.) join together as the members of the living whole which is the "total fertility" or the "total life-capacity" of a country or of an entire continent. All these members and organs are in inner relationship, each belonging to and depending on the other. If the organic unity is altered at any point, this means, at the same time, a change in all other organs. A level of maximum effectiveness depends upon the working together of all these factors. To strive to attain this effectiveness ought to be the task and aim of every agricultural measure.

The fundamental conditions of life in Central Europe permit the use of the land up to its maximum capacity. Favourable factors consist of the climatic situation and the distribution of hilly land and mountains, lakes, moorland, forests, and arable areas. Even the North German plain is broken up "at just the right point" by mountain ranges. However, a diminution of the present forested area by even 15 per cent or thereabouts would lower the biological "critical point" of effectiveness and bring about new climatic conditions which in the course of the next few centuries would produce effects tending toward the development of a climate similar to that of the Russian steppes.[1]

Considered from the aspect of a general land hygiene, of soil use and the maintenance of soil moisture, moorlands and swamps may perhaps seem unnecessary, even harmful. But do we realize that a boggy moor, under certain circumstances, can be the source of the moisture so essential for forming the dew in the surrounding plain in dry times? The drying up of such marshes means a lessening of the dew formation just at the time when it is most needed. Drain-

---

[1] In an address given before the Austrian Lumber Commission (1937) it was pointed out that the world's stocks of the Conifer, i.e. the stocks now available for commercial purposes, will be exhausted in about forty years. This was true if the devastation of the forests continued at the present rate; and World War II has increased this rate. Because the use of wood for industrial purposes of many kinds, e.g. paper, artificial silk, cellulose, artificial resin, etc. is increasing enormously, we are faced with the danger of a much more rapid disappearance of the wood supply. The problem of an increase in the formation of steppes is not only a well-known danger for the United States of America, but the subject has in recent years been discussed seriously in Europe because of the migration of the Russian steppe plants into Czechoslovakia and even into the most fertile lands of Southern Germany.

age operations, it is true, may add a few extra square miles of land for cultivation. But by this process there is the risk of lowering the value of a much larger surrounding area which may be in the highest state of cultivation. Naturally, in these remarks nothing is said against an intelligent handling of swampy regions for the purpose of fighting malaria and other diseases. The essential point is that in every case an adequate amount of water should be allowed to remain for the purpose of evaporation. But this water can be kept flowing by proper regulation and need not stagnate.

There is another, apparently unimportant, sort of local change, the consequences of which I have been able personally to observe. Numerous elms were planted during the last century in Holland. A monoculture of this beautiful and stately tree covers the flat lands along the sea coast. This growth does not so much constitute woods as merely an interruption of the plain by groups and lines of trees. Besides the lovely landscape they create, these trees form a valuable screen against the wind which at certain times blows in from the sea with great force, chilling and drying up the soil.

In recent years, however, the Dutch elm disease has to a large extent wiped out this fine stand of trees. Now the winds blow unhindered over wide stretches of the plain. My own practical experience in this connection has demonstrated that it has now become impossible to replant such a fallen line of trees. This is so because between the fallen row of elms and the shore dunes, about two miles beyond, other trees which had grown in lines and clumps have also fallen. The pressure of the wind can now come in unhindered from the sea, so that the second growth can no longer withstand it alone. The result has been that new young roots were torn to pieces and dried out. Even where a tree here and there did manage to reach a height of more than nineteen or twenty feet, it was without fail eventually broken down by the gales, and this in a locality where once there stood a mighty row of old elms.

Tree planting in this locality would be possible only by proceeding slowly and with care. Starting at the coast line, protective vegetation would have to be built up. Under the shelter of the dunes and the bushes growing on them, young shoots might get a start. After they had reached a certain stage of growth, they, in turn, would serve as protection for another line of trees planted farther inshore. The accomplishment of such a plan, however, would need the co-operation of neighbouring landowners. Unfortunately intelligent vision for this is lacking. Consequently, under the strong pressure of the wind the top soil is drying out, the

SOIL FERTILITY, RENEWAL AND PRESERVATION

ground is becoming chilled and the water table is being lowered.

One landscape architect with whom I discussed this problem felt that what had happened was rather to be desired—*that the flat character of Holland was now coming into its own.* He believed there was an aesthetic advantage in the absence of hedges and trees. Poor Holland, whose very name—Houtland or Woodland—might perhaps indicate the existence of a countryside of a very different character in the past! Surely the fruitfulness of the land is the important thing, and hedges are the most essential factor in this connection. Hedges grown to heights of six or seven feet can keep wind from the ground for a distance of nearly 500 feet and raise the ground temperature from one to two degrees centigrade to a distance of 300 feet. The catastrophe of World War II has increased the dangers for treeless Holland in many and various ways. For the biology of this poor country the loss of tree shelter may become much more of a problem than the salt content of the flooded areas.

Contributions to the farmer's soil come from a distance in the form of dust. There are many natural sources of dust. One of these is volcanic eruptions. The famous eruption of Krakatao ejected dust into the air that travelled around the earth for several years as clouds, and could, even thirty years later, be detected in the atmosphere. The ashes of Vesuvius which buried Pompeii to a depth of several feet are a classic example. They produced an especially fertile soil. Fine ash from that eruption was carried as far as Asia Minor. Only a few years ago a volcanic eruption in South America emitted ash which, perceptibly darkening the atmosphere, reached even as far as Europe. It has been calculated that the fine, invisible dust reaching Europe from the Sahara amounts in a hundred years to a layer 5 mm. thick. In the Middle Western States of America in the last hundred years, about one inch of soil has been blown from the western prairies to the Mississippi basin. Great dust storms recurring annually deposited in the basin 850 million tons of dust per year from a distance of over 1250 miles, even before the disastrous dust storms of recent years.

The famous Sahara dust storm of February 1901, which affected an area of about 170,000 sq. miles, brought to Europe about two million tons of material, and distributed 1,600,000 tons in Africa. A year later another from the Canary Islands brought about ten million tons to England and Western Europe. Such dust is either brought down by rain or it adheres in passing to the moist ground and vegetation. In China the wind-blown deposits reach a depth of several hundred feet. These deposits help produce very fertile soils.

THE FARM IN ITS WIDER CONNECTIONS

Curiously enough, the dust deposits of the American Middle West have a composition similar to loess (diluvial deposits of loam). Passarge states that the black earths of Southern Russia are the best and most typical development of loess, for loess favours the development of steppe flora and, on account of its porous character, is easily impregnated with humus materials.[1] It would seem that the black earths of Morocco are also a loess formation. According to Th. Fischer, southerly winds from the steppe lands west of the Atlas Mountains bring dust which is retained by the vegetation and precipitated especially by the very heavy dew at night. Dust deposits and decaying vegetation are thus the origin of the extremely fertile black earths.

With a content of from 25 to 34 per cent organic matter and a remarkably high potash content of from 1 to 4 per cent and even 6.9 per cent in volcanic ash, these deposits from the atmosphere are a well-balanced fertilizer. Volcanic ash is the cause of the fertility of the plantation lands of Central America and the East Indies. The Chinese peasant welcomes dust-laden winds in the Yangtze basin as bringers of fertility. So did the farmer in the prairies and South African veldt until the catastrophes of recent years. Pointing out the importance of the quality of atmospheric dust deposited annually, Dr. Treitz sums up his conclusions by saying:

(1) Dust deposits restore the exhausted bases and soluble salts (calcium, iron, potassium, etc.) in the soil.

(2) Inoculating substances for stimulating life in the soil are carried in the atmospheric dust.[2]

Braun, Blanquet and Jenny have shown that the soil of Alpine meadows is largely formed by dust particles from the neighbouring mountains; deposits of 2.8 to 3.7 pounds per sq. yard per year were measured. Discussing the nature of peat ash, which besides a considerable water content also contains mineral ingredients—especially argillaceous sand, magnesia, gypsum, oxide of iron, in addition to some alkalis, phosphoric acid and chlorine—the same authorities state that at least a part of these substances could have been brought into the pure peat-moss only by the agency of wind. The quantity is not inconsiderable since dry peat-moss contains an average of 10 per cent mineral substance.

Heavy deposits are also made through the medium of the air in

---

[1] These facts are taken from "Die Wirkung des Windes" by S. Passarge, *Handbuch der Bodenlehre*, Bd. 1, 1929, Berlin.

[2] From the report contributed by W. Meigen, "Material aus der Atmosphaere, *Handbuch der Bodenlehre*, Bd. 1, 1929, Berlin.

the neighbourhood of factory towns. It is calculated that 224 to 296 pounds of soot and dust per acre per month fall on the ground in industrial areas. A third of this deposit consists of soot and sulphuric acid. Additional fertilizing substances are also present in the rain-water received by the earth. For example, rain-water contains the important fertilizing element carbonic acid, of which over a million tons fall yearly in rain-water in Germany alone. On the sea coast there are the salts brought by wind and rain. Then small but active quantities of iodine—0.1 to 0.2 cubic mm. per 100 cubic yards—are present in the air. There is also chlorine. Along the English coast, one quart of rain-water carries at least 55 milligrammes of chlorine and even inland 2.2 milligrammes are deposited. During storms in Holland, 350 to 500 milligrammes per quart of rain-water were measured. This means that we have in these regions a chloride fertilizing from the air ranging from 1 to 13 pounds per acre. In many tropical regions such as Ceylon this figure rises to 30, to 50, or even 150 pounds per acre due to the strong evaporation of the sea-water.

It is also surprising to find that other substances important to plant nutrition are present in measurable quantities in rain-water. There is a particular relationship between nitric acid together with other nitrogen compounds and rain, and between phosphoric acid and snow. This is not without importance for the practical agriculturist. In his desire to conserve the nutrients in the soil and to set free those bound up in it, he often resorts to one-sided chemical fertilizing, not realizing what assistance Nature offers of her own accord. Nitrogenous compounds rain down on England for instance at the rate of $3\frac{1}{2}$ pounds per acre, and in Northern France 9 pounds per acre, per year. Measurements of sulphuric acid yielded the following figures per acre: in Giessen, 8 to 100 pounds; in Cologne, 200 to 340 pounds; in Duisburg, 200 to 600 pounds per year—unquestionably a considerable fertilization.

One more source of material from the atmosphere should be mentioned and that is the so-called cosmic dust and meteorites. A significant quantity of such cosmic substances reaches the earth throughout the year. A continual change of substance takes place between the earth and cosmic space. In the case of meteors the masses are concretely perceptible since individual specimens often weigh many tons. And to this must be added the shattered dust-like particles of shooting stars. The quantity of this cosmic dust varies between ten thousand and one million tons per year. It even causes the formation of certain red deep-sea sediments.

THE FARM IN ITS WIDER CONNECTIONS

The mention of these facts is not intended to imply that everything needed by the farmer comes from the air and that therefore manuring is unimportant. *Good manuring is always the basis of all agriculture.* These facts are only cited to point out that there are often active factors in addition to those involved in the nutrition equation. There are more processes interacting between soil and plant in Nature than in a test-tube. An extension of our knowledge in this direction is a vital necessity. We should learn to understand and be able to reckon with the biological processes in Nature.

Mineral substances are collected by the plants themselves and thus brought to the soil. This will be discussed later in connection with plant nutrition.

CHAPTER FIVE

The Soil, a Living Organism; The "Load Limit" in Agriculture

The soil contains dissolved and undissolved mineral ingredients, water, organic matter from living plants, and also organic and inorganic substances that have originated from the decay of roots or entire plants. Furthermore, it contains such living organisms as bacteria, earthworms, insect grubs, and sometimes even higher animals, whose activity and decay contribute to the transformation of the soil, physically by breaking it up, and chemically by feeding, digesting, etc. The interaction of *all* these factors together with climate, daily and annual weather conditions, and the cultivation that the human being gives it, determine the fertility of the soil.

The purely mineral, undisintegrated rock of the high mountains is infertile. If its surface is decomposed and made porous through the action of the heat, frost and rain, it becomes possible for the first beginnings of life to appear within in. Weathering gradually creates a crumbly top layer, a soil surface. The more intensive and rapid the weathering process, the more fertile becomes this layer. This cyclic weathering process repeats itself year after year and acts upon newly disintegrated rock as well as on every piece of soil already tilled. Through frost and rain, a physical disintegration of the soil takes place. In summer, through the influence of warmth, radiation and the life processes in the soil (bacteria, etc.), a dissolution of a more chemical nature occurs. The weathering process may reach a point where the various constituents separate, everything soluble being carried off by water. Sand and clay form the chief mass of *alluvial sands* in contradistinction to *weathered soils* which include all kinds of mineral substances such as silica, volcanic rock, or chalky marl. The individual rocks yield substances susceptible to weathering; thus volcanic land, as a result of the igneous process it has undergone, is constituted differently, for example, from northern mica schists. The former disintegrates rapidly with fertile results, the latter only reluctantly, giving a sparse soil. Limestone is especially resistant to weathering, and only covers itself with a thin layer of marl. In more southern regions, the chemical weathering, especially oxidation, is more intensive under the influence of the sun's radiation. There it is often lack of water that keeps the ground from becoming fertile. Thus every soil, in accordance with its geological ancestry and age, represents a particular state of weathering. The organic processes co-operate, the disintegrating plant and animal bodies eventually furnishing humus.

The quality and amount of the humus determine the fertility of the soil. Not all organic constituents of the soil need necessarily be humus or even potential humus. Only about 40 per cent of the organic constituents are humus or humus-like substances. The actual content of neutral colloidal humus substances is, even in a soil rich in organic matter, relatively low, and generally less than 1 per cent in the case of highly cultivated agricultural land in the Temperate Zone. Just as the inorganic ingredients of the soils may weather differently, according to climatic conditions and the sort of cultivation given to the soil, so there is even a greater difference and wider variation in the organic parts. The most important of the various factors governing decomposition are intensive aeration of the soil, long drought, heat, and intense solar radiation, which break down the substance into ammonia or nitrogen and water. This means that these things are lost as fertilizers for agriculture. A hard soil surface, moisture, cold, continued cloudy weather, accumulation of water in the soil, insufficient drainage bring about a humus material rich in carbonic substance and poor in oxygen. Such soil gradually grows more and more sour and, if air is kept away from it completely, it eventually turns to peat. Such soil is indeed rich in organic substance, but this material is in a sour, matted state and hence not directly usable. The valuable neutral colloidal humus is lacking. Organic substance of this kind is a dead mass as far as agriculture is concerned.

Special attention is called here to the fact that there is no exact chemical formula for humus. All we know is that in humus we have a mixture of the products of disintegration, more or less rich in carbon, nitrogen and oxygen. In humus the essential factor is its *condition*. The degree of organic disintegration, together with the state of the soil, chemical and physical, determines its value.

Soil bacteria play an important part in humus formation; these in their digestive activity, and by means of their dead remains, are important contributors. In rich and healthy soils they furnish as

SOIL FERTILITY, RENEWAL AND PRESERVATION

much as 600 pounds of humus per acre—a considerable fertilization. This is of even greater value inasmuch as the bacteria are gatherers of phosphoric acid, their bodies being rich in this substance. They also assist in setting free phosphoric acid already present in the soil. Phosphoric acid, as is well known, is usually present only in very small amounts in an available free state. It is the task of these minute organisms to help quicken the soil by liberating its natural reserves, thus adding to the phosphoric acid in manure and compost. Besides this—according to the researches of the Swiss chemist Paulus—the atmosphere itself holds finely diluted quantities of phosphoric acid. In winter, especially in snow, measurable quantities of it are deposited.

The process of nitrogen-fixation by bacteria is well known. May it not be possible that a similar process exists in the case of phosphoric acid? This is a question which deserves investigation and research.

The author is acquainted with the studies of noted research institutions which report that on large test areas no diminution of phosphoric acid has occurred despite intensive cultivation of the land; although no phosphoric acid was added other than that provided by the normal manuring of the land. It was observed that the P_2O_5 content of the same soil was subject to fluctuations at different times of the year—an indication that one is dealing here with organic processes within the microscopic world of the soil.[1]

Hence the problem of phosphoric acid appears not to be a part of inorganic soil science, but something that is essentially connected with the whole humus problem, and therefore worth referring to at this point.

The most important humus-maker in cultivated land in the Temperate Zone is the *earthworm*. It digests organic refuse combined

---

[1] The degree of acidity of a specific soil varies with the seasons, showing a minimum in the autumn months and a maximum in the cold winter months, with 5 further secondary minimum and maximum during the year. There is a further variation depending on the natural sour or neutral condition of the soil. Sour soils have a minimum in June and a maximum in March; neutral soils, a minimum in May and October and a maximum in February and August. It will therefore be necessary in the future to know exactly when soil analyses are made, for they can be compared only with other tests made at a similar time of year. Periodic variations with two minima and maxima have also been detected in the humus and nitrogen content of soil; a cyclic curve for phosphoric acid and its solubility in forest soil has also been observed showing a minimum in summer with a maximum solubility in autumn and winter. According to Prof. Feher of Sopron, phosphoric acid analyses without any indication of the time of year are of no practical value in view of the seasonal variations of its soluble or insoluble compounds.

with the mineral components of the soil, and then excretes humus. It also aids soil drainage and aeration by making small holes and passageways. In virgin soil, humus gives natural fertility and a light, crumbly structure even to heavy soils; light soils are also protected against drying out and against erosion, by a capacity to hold water that they derive from the humus substances excreted by the earthworm. It is not astonishing that Charles Darwin in 1881 dedicated a whole book to the fertilizing activity of the earthworm. He says that without this worm there would be no soil for agriculture. Since that time many studies of the earthworm have been made. It is estimated that the soil of a meadow in good condition contains from 200 to 300 pounds of worms to the acre. For the farmer the presence of a large number of earthworms in the soil is the visible sign of natural biological activity. The use of the microscope is not practicable for him, so he must make use of this natural barometer of fertility.

Every measure, which disturbs life in the soil and drives away the earthworms and bacteria, makes the soil more lifeless and more incapable of supporting plant life. In this connection we see the dangers of one-sided fertilizing, especially when one uses strong doses of chemical fertilizers containing soluble salts like potassium or ammonium sulphate, or highly corrosive substances such as nitro-phosphates (usually under some fancy trade name), or poisonous sprays, such as arsenic and lead preparations. These injure and destroy the micro-organic world. Soils intensively treated with chemical fertilizer or orchards sprayed for a long time with chemicals have no longer any biological activity. The author has seen vineyards which had been treated for years with copper and lime solutions absolutely devoid of earthworms—that is, devoid of the creation of new humus.

Once the humus content is exhausted, new conditions arise in the soil; the mineralized structure then resembles more the laboratory conditions of a purely chemical relationship between soil and plant. Then the law is completely valid which says: "*What is taken from the soil in mineral substances by harvest must be returned to it.*" This sentence is true as far as laboratory research is concerned. In a closed system made within the closed containers of a laboratory, this rule was discovered and established.

This same rule would, indeed, hold good in nature too, were Nature herself composed only of lifeless mineral ingredients. Through our modern intensive methods, especially in our use of strong or exclusively chemical fertilizers, we have created such conditions that the purely physico-chemical qualities of the soil are made predominant while the organic activities are eclipsed. The mineralizing of the soil becomes visible with the disappearance of earthworms and with the formation of a surface crust which can be lifted off like a shell in dry weather. The latter phenomenon ought to be regarded by the farmer as a "gale warning" given by his "soil barometer".

When the winter furrow, weathered by frost and rain, shows a porous, uneven and irregular surface in moist days in the spring, or when harrowed land still remains porous, light and crumbly after rain, or when in summer, after a long dry period, it shows little or only a thin skin and the earth dries in fine cracks, then biological activity is present. Soils of this type may be called elastic, but not only because of the springy, elastic feeling one has when walking across them. This term elastic is still more descriptive of the relatively resistant powers of such soils against all sorts of harmful influences—for instance, mistakes in crop rotation, or in tilling or harrowing when the soil is too wet. Above all it is elastic in its retentive capacities with respect to water. Water, which is not only important for plant nourishment, but also makes up some 40 to 80 per cent or more of plant substances, should under all circumstances be retained by the soil. An important sign of a living soil is its capacity for absorbing water. When the soil is alive, a sudden summer shower is immediately absorbed and disappears as if into a sponge. When puddles remain on the ground in wet weather, despite the fact that drainage is otherwise functioning well, it is a sign that conditions are not good.

A most impressive example of such conditions was seen by the author on a visit to the Agricultural College of the Missouri State University. A large experimental plot had been divided into two parts. During thirty years one part was left untouched as virgin prairie; the other had been tilled for three decades after the manner of modern intensive cultivation. As a result, the virgin prairie, although a heavy, dark brown clay, was porous, so that a stick could be driven into it to a depth of more than eighteen inches. Rain-water disappeared immediately. Even four years after that visit, the following experiment could still be made on samples of this soil taken away and preserved in the author's laboratories.

If some of the untilled soil is placed in a test-tube and covered with water, the water is immediately absorbed, and a damp-dry mass is the result. After drying it out, the test can be repeated at will.

On the tilled half of the field in Missouri the surface of the earth had become harder and harder and a stick would penetrate only a few inches into the ground. Rain-water stood in the furrows and it ran off wherever there was a slope, making channels, the beginning of dreaded erosion. With such mineralized soils the danger of erosion from rains, accompanied by a drying-out process, is very great. Every good farmer living in hilly country and in hot climates knows this.

The mineralization of the soil can be caused by the destruction of the physical and organic structure. We shall speak later of the influence of cultivation. The effect of the use of chemical fertilizers on plant growth is well known. They increase the yield, producing thick, heavy plants, especially when nitrogenous fertilizers are used. This effect has made the farmer particularly fond of chemical fertilizers, for they give him bigger harvests. Scientists even tell him that he thus supplies otherwise deficient elements in the soil.

But two things may be observed by the practical agriculturist who lives and works close to the soil:

(a) That in order to maintain the same yield from year to year he must increase frequently the yearly quantity of chemical fertilizer.

(b) That the soil structure changes in the direction of the hardening and encrusting process already described.

Why do agricultural schools and research stations remain silent concerning these phenomena which are familiar to a large percentage of practical agriculturists? A great deal is said concerning the value of artificial fertilizers for increasing yields, *but very little about the alteration in the soil.* Perhaps it is because the soil used for parallel tests in experimental stations is already in such a condition that this alteration in soil structure is no longer obvious.

Occasionally we find in scientific literature a note to the effect that chemical fertilizers must be given in the correct proportions in accordance with soil analyses. But since, in most cases, a field is not uniform in soil content, we should, theoretically, have to use a different formula for each of its variations. That would be particularly difficult in soils that have been produced by weathering. As experts in the science of fertilizing themselves readily admit, this differentiation is, indeed, theoretically necessary, but impossible of practical application. Therefore they take an average formula and accept unbalanced fertilizings as inevitable in practice. Another point to be considered here is that not every farm manager is in the position to have an adviser standing at his elbow who is experienced in the problem of *his particular soil.*

In regions where enough moisture in the form of atmospheric humidity, sub-surface water and rain is available, fewer difficulties are encountered at first in respect of hardening and the alteration of the soil structure. This is because the normal admissible balance of mineral salts is again produced through the balancing, distributing and dissolving effect of the water.

In Holland, for example, the bad effects of artificial fertilizers appear much more slowly, because the high level and the constant circulation of the water table in Holland provide for a natural balance. But what happens in a dry summer or in an arid climate? In addition to the encrusting of the surface, another phenomenon may be observed. In a field will suddenly appear small, less fertile patches. These grow from year to year, and in dry years there will be more and more of these sterile spots. We may ascribe these to so-called free acid products, that is, insoluble or practically insoluble silicates which are formed out of the soil silicates in the interchange of salts. These insoluble silicates must be regarded as lost to the substance cycle of agriculture, and can only be won back after a long organic "cure".

In order to explain the foregoing, it is necessary to remember that, according to physico-chemical laws, a state of equilibrium exists among the various salts in a solution. If one adds an easily soluble salt to a mixture of salts already in solution, the less easily soluble can thereby be forced out by precipitation.[1] This process is widely used in the chemical industry, in the precipitation, for example, of organic dye-stuffs by means of sodium chloride, ordinary table salt. In the soil there is a mixture of salts: silicates that are dissolved with difficulty, salts and silicates easily washed out by weathering, and easily soluble salts—especially potassium salts—and salts arising from organic ingredients, etc. When the easily soluble fertilizer salts are added, the balance turns in their favour against the heavy silicates. That is, the weathering and disintegration of the ingredients of the soil are slowed up and in most cases are made useless. The plants *are* then in fact dependent on the soluble fertilizer salts and they react clearly to their dependence and their poverty. Under such conditions, the *soil chemist* is indeed correct.

The *soil biologist* starts from other premises. For him the earth represents a natural reservoir, and, except where pure sand or undisintegrated stone is present without water, a more or less complicated mixture of salts, silicates, zeolites, aluminates, etc., will always be available. These mixtures represent the natural basis when it is made available for the substance cycle. *We should not precipitate the salts in our soil*, but should rather open the soil and en-

[1] Gibbs's law.

liven it. The organic processes taking place in the soil, due to light, air, weather, microscopic organisms and humus, are the activities that help its "demineralization" on the one side, and foster its "becoming organic" on the other. In so doing they open up extensive reserves. When, as a result of a single year's harvest, the soil substances are used up, new substances are again set free through the natural action of weathering in a living, vital soil. Tillage and the eroding effect of wind and water liberate new soil ingredients every year. These need only to be enriched, "organized" in the most literal sense of the word.

In Central Europe we estimate a yearly loss of 1 millimetre from the surface of a field through action of wind, rain and tillage, while in alluvial lands there is a constant addition of material. For example, 1 millimetre of soil per year is added in the Nile Delta where there is a wonderful natural irrigation.

The yearly ploughing and harrowing of a field accompanied by the loss of 1 millimetre from the surface results in the fact that each year the depth of the cultivated portion of the soil moves 1 millimetre lower. In this new millimetre a reserve mass of mineral ingredients is uncovered. The essential thing is that this reserve be really liberated and brought into the cycle of agriculture. With a yearly deepening of only 1 millimetre of soil, about 6·4 tons per acre of new soil are brought into action; this constitutes, even in very poor instances, a valuable reserve. This soil includes potassium and above all phosphoric acid which in normal soil is transformed by micro-organisms, but which by means of various salts such as potassium is in danger of being fixed in insoluble forms from which no good can come.

In studying such relationships we see an important difference between the conditions in the research laboratory and those in nature. If in addition we count on accidental sources of mineral substances during the course of the year, especially on those brought by wind and rain, we can then see *how little the cultivated field constitutes a closed system of substances*, if we are only careful in our methods of cultivation to preserve the living elements of this nature, and not let them be precipitated. Among the occasional sources of additional substances should be included the salt content of rain and snow as well as the movement of dust of both organic and inorganic origin.

By observing the varied colours of the winter furrows at the close of several winters, we have a good example of how differently the weathering processes work on the decomposition and structure of the soil in different years. When we take the above into consideration and with it the fact that plant life, by having its roots in the earth, sets up a whole series of processes and transformations in the soil, it becomes quite clear to us that *the cultivated field is a living organism, a living entity in the totality of its processes.*

THE "LOAD LIMIT"[1] IN AGRICULTURE

The scientific farmer of to-day talks continually about high production. And this high production at which he aims obviously depends on the general capacity of his farm. This capacity, in turn, depends on a variety of factors connected with the life process previously discussed. Therefore, his first task should be to carry on his work in accordance with a complete knowledge of the nature of this life process, just as, in another field, the engineer bases his own work on a careful study of the strength of his materials.

Why are we always saying "High production! High production!" without at the same time considering the use of the construction engineer's mode of thought, which has as its basis a real knowledge and an understanding of the "materials" dealt with? Corresponding to the "load limit" in the inorganic, there is, equally, a limit of inner capacity, a biological "elastic limit", a biological "resistance point",[1] for the "living material"—that is to say, for the organism in the widest sense of the word. If the limits of these expansion values are overstepped, harm will come to the life phenomenon just as the collapse of the mechanical structure occurs if the resistance point is overstepped.

A cultivated field has not got an unlimited capacity for increased productivity. Its productive capacity is not directly proportional to the amount of fertilizer applied. A cultivated field is also a biological organism, and as such is subject to the laws governing the organic. *It has its critical point of inner effective power.* This is defined as the product of many varied factors. Among these are: the mineral constituents, the physical structure, the presence of organic substances and their conditions (humus, acidity, etc.); climate, methods of cultivation, varieties of plant growth, activity of plant roots, ground cover and cover crops versus erosion, possibility of the occurrence of weathering processes, proximity or absence of woodland, etc. All these factors together determine, in the first place, the capacity for biological activity. If any one of these factors is disproportionately active, a disturbance arises, and

---

[1] These expressions are deliberately used in the sense of the mathematical-physical observation of nature.

in time a weakening of the entire system occurs. By strengthened fertilizing, an increase in production is obtained; but if this is overdone (in a one-sided way), the soil as an organic structure goes to pieces. The basic principle underlying a healthy, *lastingly productive* agriculture is, therefore, *knowledge of the "resistance point" of a specific soil.* This alone gives assurance of lastingly healthy conditions, making productive capacity certain.

As a living entity the soil's capacities have a definite limit. This is also true in the case of a machine. It ceases to run when it is overloaded, it breaks down, becomes quite unusable, but it does not restore itself. A living entity will bear a certain amount of overburdening. For a short time, more can be demanded of a draught animal than is good for it to give; it will recover its strength again. Only through long and continual misuse is it weakened and its offspring made less valuable. This weakness, however, is not immediately noticed. When it finally shows itself, it is frequently already too late to remedy the error. It is easier for the keen observer to prevent a sickness than to heal the already attacked and partly or wholly injured organism. This is also the case with the soil; it has a natural standard of life. To recognize and guard this is the greatest art of the farmer. If an individual field is expected to produce out of proportion to its capacity, it can—especially by increased fertilizing—be whipped into carrying an overload for a certain length of time. Then however troubles begin, and the longer these last the more difficult are they to eradicate or to cure. It is then of the greatest importance for the practical farmer that he have a correct understanding of the *natural load capacity of his soil.* Only when he really has the same regard for his soil that he has for his horse, the welfare of which he guards daily, can he expect to get from it a performance commensurate with its capacities, year in and year out, without detriment to it. Everything that he does in connection with his soil must be regarded from this point of view: ploughing, harrowing, choice of seed, as well as manuring.

Neither soil nor cattle nor the human being himself are, in respect of the manifestations of their life, an arithmetical problem. Every jockey knows that the performance of his horse, its pace, its jump, its endurance, does not depend on feeding alone. This obviously creates the physical basis, but the horse's protein or lime content does not transform itself proportionately into jumping ability. Nor are heavily fed horses the ones with endurance. The rider knows that there is much besides feeding taking place between him and his horse which has an effect on calling forth its best per-

SOIL FERTILITY, RENEWAL AND PRESERVATION

formance. Unfortunately, the farmer does not regard his cow in the same way. She has already largely become a subject of arithmetical calculation. In one end are stuffed so and so many pounds of proteins and salts, in order that so and so many gallons of milk may flow out of the other. The cow co-operates for a while, then we are amazed at weak calves, streptococcic diseases, abortions etc., phenomena that arise "inexplicably"!

A cow by intensive feeding can be brought to maximum yields of milk. These represent the result of a one-sided performance on the part of the organism. Milk production is a part of the animal's sexual activity. An intensification on the one side means a lessening on the other, in this case a weakening of the organs of reproduction. We thus get a higher milk production; but on this account we also have to deal with all the familiar difficulties of breeding and raising cattle, such as contagious abortion (Bang's disease), delayed and heavy births, sterility, mastitis and other streptococcic infections. The "expansion limit" of the productive capacity of the cow is a product of breed, constitution, size, quality and quantity of feeding, individual capacity for the utilization of food, pasturing or stable feeding, local terrain, climate, ability of the breeder, etc.

We have become even more accustomed to regard the soil as a real equation of nutritive values. Such an equation would, as a matter of fact, be correct, if we included *all* the factors. But the following is an example of one-sidedness and an improper equation: soil plus additional fertilizers equals soil plus yield.

The proper equation, however, from the point of view of life, should be as follows:

$$\text{Natural fertility plus production capacity equals function of} \begin{cases} \text{soil} \\ \text{manuring} \\ \text{humus} \\ \text{tillage} \\ \text{rotation} \\ \text{climate} \\ \text{weather conditions} \\ \text{quality of seed} \\ \text{weed growth} \\ \text{and a number of environmental factors} \end{cases}$$

The consideration or neglect of *any one* of these factors is just as important as the whole fertilizer equation. What then can the farmer do in order to be fair in all these, and maintain at a high level his living organism, *the farm?*

CHAPTER SIX

The Treatment of Manure and Compost

The first problem facing the farmer is the proper use of manure and the correct treatment of compost. Contrary to the general popular belief, *we feed the soil by manuring*. We do not feed the plants. The vital activity of the soil must be maintained. Nature takes care of this by humus-developing activities, bacteria, earthworms, roots that break up the soil, and weathering. The farmer's first task is to aid these natural *organic* processes in the soil. When manure is applied it must enter the soil in such a condition as to contribute to this work. It is least able to do this when it is raw and fresh. Unrotted manure, in its process of decomposition, actually feeds on the soil for a certain length of time. This is so because unrotted manure requires biological activity and energy in order to rot. Especially objectionable is the absorption of the decaying products of half-rotted albumen which may be taken up directly by the plant roots. This can have a disturbing effect on plant and also on human health. In the familiar smell of cooking cauliflower, the sort of manuring treatment which the plant has received can be detected. *One can actually smell in the kitchen the pig manure, the sewage, etc., which have been worked into the garden soil.*

The best form of organic fertilizer is humus. Unfortunately a considerable length of time is required for stable manure to rot and become humus. And during this rotting period valuable substances are disintegrated and lost, if leakage is not checked. Generally speaking, stable manure is exposed to an endless series of losses. In the first place there is the loss of nitrogen under the influence of certain bacteria and the weather. These especially rob the surface of the manure mass wherever too much air is allowed to penetrate. The usual method of throwing stable manure loosely out into the manure yard, and exposing it to the sun and rain, may cause a loss of 50 per cent and more of nutritive substances. Some of its nutrition is dried out by the sun and oxidized away and more is washed out by the rain. If the manure stands on a slope, one sees a brown stream running out of it, carrying away the most important nutritive substances. The same farmer who has carefully calculated from fertilizer tables the amount of ammonium sulphate he needs in order to have a profitable harvest will, curiously enough, watch with the utmost tranquillity the nitrogenous substances of his manure running away in liquid form down the road, fertilizing the pavement and drains or, at best, the weeds along the ditch. I once met a man, exceptionally "practical" in his own estimation. Below his manure heap there was a small pond to which the brown liquid from his manure heap was running in numerous rivulets. *"Oh, that is all caught in the pond,"* he said. *"We clean it out every four years!"*

Then there is the opposite sort of manure heap, one that gathers all its liquid at its base; this slowly wets the base and then rises higher into the body of the heap. This condition quite insulates the part standing in the water or liquid manure from the air. Thereby, a proper fermentation is prevented, with the result that instead of a good fertilizer material we have a black, strange-smelling mass, a substance turning into a sort of a peat. The value of this as manure is slight. Such a product is more like a wet loam or clay than manure. It smears the ground, is taken up with difficulty, and even after weeks the black lumps are to be found in the soil still almost unaltered. Processes have taken place in this case which go far beyond the goal of humus formation, in contrast with manure which is dried out by too much exposure to the air. Manure which is, on the other hand, too tightly packed runs the danger of heating too much and of losing its best qualities. In short, there are many wrong ways of treating manure which may result in the loss of half its original nutritive value. Having calmly thrown away a large part of the nutritive substances in his manure, the farmer rushes to make up his loss by the use of mineral fertilizers. Even so, he soon discovers that these are only a temporary help.

The first and foremost practical rule for the right treatment of manure is to heap it up daily with proper care in the manure yard. It is best to start in one corner of the yard and build a firmly trodden, but not stamped-down, rectangle from two to three feet high, with a base area of two or four square yards. Then a second heap is placed alongside it, then a third, and so on. The first section may be covered temporarily with a few planks, so that a wheelbarrow can easily be run over it to the next. It is of the greatest importance, however, that the manure be always well covered. It is only when the entry of air into the heap is lessened by proper covering that the action of the bacteria which rob the surface is arrested. The best covering for the heaps is earth. With medium soil a four-inch covering is enough; if the soil is heavy less is needed. The leaching and drying-out processes are prevented by covering. Only where no soil is available should the covering be of peat-moss, planks, straw, or potato-plant thatch. If potato plants are used, it must be realized that a certain amount of their own valuable substance will be lost as they rot on the surface.

The thickness of the earth layer, as already indicated, depends very much upon the kind of soil used. However, the contents of the heap should on no account be *completely* shut off from outside influences. Impermeable soils, such as heavy loam and clay, must

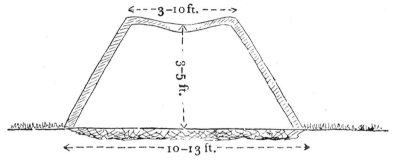

FIGURE I. MANURE HEAP AND COVERING.

be applied thinly, especially when these are put on in a wet state. If sand is used, care must be taken that it will not be blown away by the wind nor be in danger of sliding down the sides of too steep heaps. The guiding principle in all this is the fact that the manure heap itself must be treated as a living organism, because of its bacterial content and its internal fermentation. As such it must have its outer boundary, or skin, separating it from the outer environment. It should develop its "own life". The decomposition of manure should not be subject to a haphazard fermentation. The one aim of the manure heap is the production of humus. And the purpose of all useful organic decay is to produce a neutral humus. Manure brought to the soil in this state not only gives it the maximum in fertilizer value, as far as available nutrients are concerned, but it is also most helpful to the physical structure of the soil.

At this point, besides the proper and careful handling of manure, begins the bio-dynamic method of soil treatment offered by Rudolf

Steiner. Its effect on the soil can be likened to the process of making bread. Water and flour are first mixed into a dough. If this is then left standing for a certain length of time, exposed to the air, any yeast bacteria that may happen to be floating about (so-called wild yeast) will fall into this mixture and in the course of several hours or days will cause fermentation. The bread baked from such a dough is sour, bitter and hard; it is inedible. In order to produce a good edible bread, the baker must use a select strain of cultivated yeast or perhaps a leaven in order to get a quick and good fermentation.

The farmer generally treats his manure after the first method; fermentation is left to chance. His proper course should be to develop a controlled fermentation, which allows only a minimum loss of nutritive elements and causes a better humus formation. In this way he follows a controlled method, not one left to chance.

Dr. Steiner has shown that such control can be obtained by the use of certain plant preparations[1] which induce the right kind of fermentation. This is done by the use of various plants, which have always been employed also as medicinal herbs, such as: camomile, valerian, nettle, dandelion, horsetail, etc. These plants are themselves first put through a long fermentation process, buried at specific depths in the earth in close contact with certain parts of an animal organism. The process can be described by saying that through a kind of hormone influence the fermentation is guided in a definite direction.

After a number of months, these plants are actually transformed into humus-like masses. If small amounts of these preparations are inserted into a carefully piled up manure heap, the entire fermentation of the heap is given the proper tendency towards humus formation. The result is that after a short time—generally two months—the dung is turned into a blackish-brown mass, rich in humus materials.

Researches have shown that, during the rotting process, the bacterial content of such a manure heap is ten times that of one not so treated. Particularly noticeable is the presence of a great number of earthworms. Such heaps are constantly filling up with earthworms which, after their humus-forming activity is completed, die and provide an additional fertilizing substance through the decomposition of their own bodies. For the attainment of the results outlined here, a number of other points must also be considered.

---

[1] Cf. Preface.

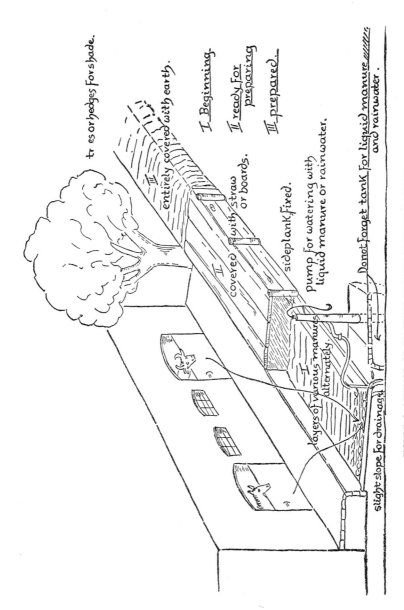

FIGURE 2. METHOD OF STACKING MANURE IN THE FARMYARD.

SOIL FERTILITY, RENEWAL AND PRESERVATION

Strawy manure, containing much air, heats easily, especially when it contains horse manure. Wet, greasy manure becomes putrid. The kind of manure produced depends also to a large extent on the feeding system in the stable. The best manure comes from a feed of roughage, grass, hay, clover, pea vines or other straw. It has the structure most favourable to fermentation, especially when, in addition, liberal use is made of straw for bedding down the animals. Turnips, turnip tops, etc., produce too wet a manure. Concentrated feeding produces a sticky manure of wet consistency which ferments slowly or not at all. The most unsatisfactory manure comes from animals fed chiefly on concentrates with little hay and when only such materials as leaves, sawdust, etc., are used as bedding instead of straw.

Any partial or complete departure from conditions which produce the best manure can easily be noted and corresponding measures taken to correct them. Because the best preservation of fertilizer values is obtained from mixed manures (horse manure especially being a protection against denitrogenizing bacteria), a mixed heap is to be recommended—but, of course, only where this is possible. Anyone who uses horse manure for hot beds must naturally store this up separately. The best practice otherwise is to take a cow manure, produced from hay and green feed and caught up in straw, and spread it carefully out over a small area in the manure yard. A thin layer of horse manure is then put directly on this underlayer of cow dung. This combination remains lying for a number of hours in order to "steam out". Then, before the next layer of manure is added on top, the under layer is trodden down until it becomes fairly firm. When the manure is strawy, this treading down can be done more vigorously; if it is too sticky and wet, it must be less solidified. If the manure originates from concentrated feeding and has little straw mixed with it, it must lie exposed to the air a little longer to "steam out", to dry out somewhat. In such a case, when possible, two piles should be started alongside each other and additions to them should be made alternately.

When the heaps have reached a height of three or four feet—strawy manure shrinks together more than other manures and can be heaped higher than wet manure, which suffers from too much pressing and packs itself too solidly together—they are then covered with earth or boards or straw. A new heap is then started or additions are made to the old one. The sides can be made of movable pegged planks, which are removed when the new batch is set up alongside. When making long heaps the procedure is to cover three

PLATE 1. A bio-dynamic manure heap

PLATE 2. Cross section of a three to four-months-old dynamic manure heap. The straw has completely disappeared. A beautiful brown mass is now available which can be spread over the field with a shovel

PLATE 3. A model compost yard

THE TREATMENT OF MANURE AND COMPOST

to four sections as soon as they are set up. Such complete sections must then have the bio-dynamic preparations[1] inserted *from all sides*, in order thus to get a properly controlled fermentation from the very start.

If for any reason the manure is too wet (this comes from the wrong kind of feeding or from too much liquid manure, lack of straw, or excessive rain), a provision must be made for drainage within the heap. The manure heap must never stand "with its feet in the water". If the manure is too wet it cakes firmly together and becomes greasy and is unable to get sufficient air for its fermentation. In a case of that sort, when the addition of straw during the piling does not help sufficiently, the required and proper aeration can be obtained by drainage.

The necessary drainage can be effected by making a core of thorn brushwood, stiff briars or brambles of the stouter sort, or by the use of perforated drain pipes. It is advisable in very bad cases of dampness, or where the heap easily gets overheated, to put in "ventilators". These can be opened or closed as need indicates. The liquid running out of the drains should be gathered in a tank located at the end of the manure yard.

A manure heap that is too dry requires watering. Dry manure gets hot very easily, and, when there is no moisture at all, a fermentation is induced which is disintegrating and destructive. The heap remains unchanged or, when some chance wetting occurs, mould may form. This whitish-grey growth, as well as the presence of wood-lice in manure and compost, is always the sign of a too dry treatment of the manure.

The correct state of moisture for manure is that of a damp sponge; no liquid should flow out of it, nor should it be stiff and dry. For good fermentation this state should be maintained under all conditions. The best means of adding moisture is through pipes inserted in the upper part of the heap. This even becomes a necessity under extremely dry conditions. But ordinarily a trough-like depression in the top of the heap is sufficient. Water or liquid manure is poured into this trough and slowly seeps into the heap. The best liquid for such use is that which drains from the manure itself, the next best is liquid manure. Pumping the natural liquid manure out of the reservoir tank or pit on to the heap once a week, in order to moisten it, offers a very special advantage, for in this way the liquid manure is absorbed by the organic matter; it does not become putrid but is drawn into the general process of fer-

---

[1] Cf. Preface.

mentation at once. In this form, the liquid manure has also an inhibiting effect on the activity of the nitrogen-destroying bacteria and thus preserves the nitrogen content of the manure. Its sharp rank effect on plant growth is in this way avoided. In many districts, particularly the hilly regions of Central Europe, it is customary to carry the fertilizing liquid manure directly out on to the pastures and hayfields. A clever peasant, whose farmyard lay high on a hill, with the pastures lower on the hillside, once built a pipe system in such a way that he only had to turn on a cock at the

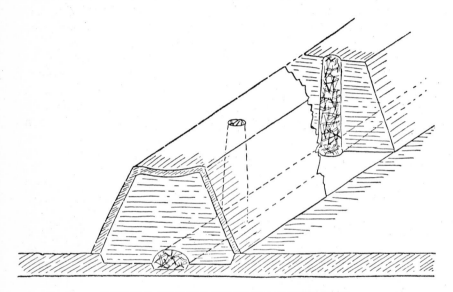

FIGURE 3. DRAINAGE SCHEME IN A MANURE HEAP.

liquid manure pit in order to accomplish his fertilizing. The harmful results of such one-sided fertilizing with liquid manure are well known. The reaction of the soil becomes more and more acid, characteristic pasture weeds appear and the growth of clover gradually ceases altogether. Furthermore, the high solubility of potash present in the liquid manure leads to a high percentage of this element in the fodder. This, as was proved by the Swiss research scientist, F. von Grünigen,[1] has in turn a baneful influence

[1] F. von Grünigen, "Mitteilungen aus dem Gebiete der Lebensmitteluntersuchungen" in *Hygiene*, vol. xxvi, Nos. 3–4, Berne, 1935.
"Die physiologische Bedeutung des Mineralstoffgehalts im Wiesenfutter mit besonderer Berücksichtigung des Kalis."

THE TREATMENT OF MANURE AND COMPOST

on the health of the animals. All such bad results can be avoided if the procedure suggested above is followed.

In general, a manure which has been carefully heaped up and given the bio-dynamic preparations becomes ripe in two months and is transformed into a humus-like mass, ready for use. If dryness or too much moisture, or an excess of one type of the manure (too much horse in proportion to cow, for example), has brought disturbances into the fermentation, such as putrefaction, mouldiness, or too much heating, then it is advisable to turn the heap. This can be done when necessary from the second month on. If after two months the fermentation has already transformed the heap into an odourless mass, turning is unnecessary. If during the turning too great a dryness is observed, together with too high a

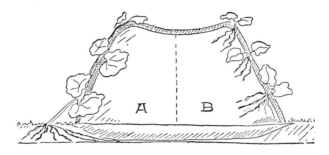

FIGURE 4. RIGHT AND WRONG WAY OF GROWING PLANTS ON A COMPOST HEAP.

temperature (anything over 150 degrees Fahrenheit must be considered harmful), then the mass must be thoroughly watered. Turning on rainy days is a good measure. Weeds should not be allowed to grow on manure or compost heaps. A growth of grass on the heaps is also harmful, for it cuts off air with its thick root system and thus checks fermentation.

On the other hand, where there is no shade the heap is exposed too much to the sun. It will then pay to screen it with matted straw, reeds, etc. If however it is necessary, for the sake of shade, to plant something on the heap, lupins or vetch are suitable or, as may be seen more frequently, cucumbers or gourds. The latter are planted in the ground around the outer edge of the heap, and the stems are trained up over it. A small hedge, or trees, should, correctly speaking, be planted around the site of the heaps. In dry summers a variation in the speed of the rotting on the north

SOIL FERTILITY, RENEWAL AND PRESERVATION

and south sides of the heaps has been noted. The shady side ferments somewhat more rapidly.

What has been said here in a general way concerning the handling of manure is equally valid for the preparation of compost. Compost is a mixture of earth along with every sort of organic refuse, which rots without having gone through an animal organism. It therefore lacks the presence of animal hormones which, even when present in infinitesimal amounts, foster plant growth.

Direct contact with the earth is very necessary for both compost and manure. Innumerable micro-organisms, which up to the present have not been isolated in pure cultures, are present in the earth, and we are therefore obliged to use the earth as the source of these organisms. For this reason it is necessary to build compost and manure heaps directly on the bare ground and to avoid anything which causes a separation, such as concrete floors, etc. In this way micro-organisms and earthworms have free access to the heap. Even grass and turf constitute a certain separation and it is therefore better to remove the turf from the site of the heap. For the same reason the compost heap, while being made, should have mixed with it soil which is "ripe", that is to say, which contains bacteria and humus. This has the effect, figuratively speaking, of a "leaven" or "sour dough" in bread making. Having developed a ripe compost, it is advisable, when taking it away, to leave a very thin layer of the old heap on the ground and to build up the new heap on this residue. Everything that will disintegrate into humus can be used for compost. All sorts of plant refuse, straw, chaff from threshing, kitchen garbage (the inorganic substances, such as broken glass, iron, etc., should be carefully removed), ditch cleanings, road scrapings, bracken, seaweed, potato plants, hedge trimmings, wood ashes, slaughter-house refuse, horn, hoof and bone meal, can be used.

The setting up of the compost heap is carried out as follows: The first step is to dig a pit for the heap from five to ten inches deep. If this pit is pure sand, then it is best to spread a thin layer of clay over the surface; at a pinch, straw may be used. This should be covered, when possible, with a thin layer of manure or compost already rotted; or if it is an old site the bottom layer of the previous heap will serve the same purpose. What has been said concerning the "drainage" and moisture of the manure heap is equally valid for compost. The structure and consistency of the compost should be moist, but not wet.

Alternate layers of compost material and earth are then laid on the heap and between these layers a thin sprinkling of unslaked lime should always be used. When the heap has reached a height of five to six feet, it should be completely covered with earth. The size of the heap should be kept to the following proportions: length, as convenient; breadth at the base, thirteen to fifteen feet; breadth at the top, six feet; height, five to six feet. Smaller heaps may be proportioned correspondingly. The dimensions given should not be exceeded. If the material is plentiful it is better to start a new heap than to exceed the dimensions given. Only in dry and hot climates are larger heaps preferable since they conserve moisture better. In wet climates narrower heaps are preferable because they permit a better air circulation. This also applies to manure heaps. The thickness of the earth—both the interlayers

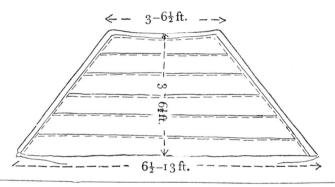

FIGURE 5. CONSTRUCTION OF A COMPOST HEAP.

and the covering—will depend on the nature of the soil in question. We should be guided by the fact that the fermentation of the heap is a life process, hence the heap must be allowed to breathe. It should have a *skin* which holds it together but which does not isolate it. If the soil used is heavy and sticky clay, the thickness should not exceed two to three inches; when the soil is light, the thickness can be from four to eight inches. It is important to note that soil from orchards and vineyards, which have been sprayed with arsenic, lead and copper preparations, is impregnated with these metals which are hostile to bacteria; such soil is entirely unsuited for preparing compost. I have seen heaps made with such soil which had not rotted after standing for two years.

It is of great importance that all compost material be moist. If, as with leaves, etc., this is not the case, the material must be moistened with water, or liquid manure must be poured on it at the

SOIL FERTILITY, RENEWAL AND PRESERVATION

time of setting up. *Maintenance of the proper moisture is one of the most important requirements of the manure or compost heap.* In time, the compost maker acquires the necessary experience. As far as possible, leaf compost should be set up immediately after the leaves have fallen and not next spring, when they will already have been "washed out". While the heap is incomplete, the material spread out on it, for example, the daily kitchen garbage, should be immediately covered with straw mats, sacking, reeds, fir or evergreen branches, or straw, etc. In tropical regions, banana leaves are particularly suitable for this purpose. If the finished heap cannot be made in the partial shade of trees and shrubs, it needs a covering which will afford a similar protection. When the heap has grown to about a yard in height, the treatment with the bio-dynamic preparations is begun. Preparations 502 to 506 are inserted at distances of about a yard apart, and the heap is sprinkled with preparation 507. What was said on page 52 concerning the treatment of manure applies here also.[1]

This treatment brings about a speeding up of the fermentation in the direction of humus formation. In about three to five months after the heap has been made it is turned, and, if necessary, the preparations are again inserted. After the turning, the material is naturally mixed together, and is no longer in layers.

The art of compost making was better known in earlier centuries than it is now even to many "experienced" agriculturists. In Flanders there was formerly a guild that had the sole right of collecting organic refuse during the daytime. This was interlaid with earth. The ripe compost found a ready market. Anyone outside this guild wishing to make compost was obliged to build his heaps secretly by night. In many places in South Wales people are still familiar with the old practice of building heaps with layers of manure, rubbish and quicklime, and covering them with earth. For such heaps it is necessary only to insert the preparations given by Dr. Steiner to add the proper "finishing touch". The "Indore process', of Sir Albert Howard based on experiences under tropical and subtropical conditions is reminiscent in many ways of the bio-dynamic treatment, except, however, in the *use* of the *preparations*.

Weeds of all kinds can also be used in the compost heap. Care must be taken that this material all reaches the inner part of the

---

[1] More detailed information on the use of bio-dynamic preparations is available through the authorized information centres and in the booklet: *Short Practical instructions in the Use of the Biological-dynamic Methods of Agriculture*, by E. Pfeiffer.

THE TREATMENT OF MANURE AND COMPOST

heap in time. Here, because of the lack of air, the conditions of fermentation are such that they destroy all seeds. In turning the heap, the outside of the original heap should be made the inner part of the new, and the former inside part now becomes the new outside. In this way the weed seeds are destroyed in all parts of the heap. Best of all is to make separate weed compost heaps and let them lie longer, even up to five months, before turning, and eighteen months before use. The normal time of rotting of well-handled compost in the damp cool climate of Central Europe is from eight to twelve months for compost material of all sorts. Cabbage stalks, which require about eighteen months, are an exception. Hence it is preferable to mix such material as this with weed compost. In tropical and subtropical climates the fermentation occurs in about three months.

The final result is a fine, aromatic compost earth, smelling like woodland soil (humus). Anyone who has much experience with the preparation of compost will soon discover two things: first, that the making of compost is an "art", second, that on every farm or garden where all the refuse material is carefully gathered, the amount of compost material is far greater than is generally believed. But this is true not only of gardening. In extensive farming also there are great reserves of compost, if we include chaff from threshing, potato plants, mud from ditches and ponds, waste straw, turnip tops not used for fodder, road scrapings and other materials. On a mixed farm of 250 acres, with fourteen dairy cows, four horses, a proportionate number of young cattle, and an intensive culture of grain, it has been ascertained empirically that the quantity of compost gathered together in a year was equal to the quantity of stable manure gathered in six months, the latter being well mixed with straw. If we include the conservation of fertilizer values resulting from bio-dynamic treatment of the organic manures and composts, we can confidently assert that by composting the fertilizing materials not used before the farmer will have, after the change of method, more than *double his previous amount of fertilizer* in respect of value and content. It should be pointed out to the agriculturist or fertilizer expert, who may think the bio-dynamic method does not adequately cover the nutrient requirements of the land, that the full value of the manure of the bio-dynamic farmer is conserved instead of being wasted and he thereby creates a reserve. He can cultivate more intensively because his organic fertilizer—the basis of all agriculture—is more intensively treated.

CHAPTER SEVEN

Maintenance of the Living Condition of the Soil by Cultivation and Organic Fertilizing

It is useless to pay great attention to the conservation of manure and compost values, if, later on, in the care of the soil, all the valuable substances are simply to be squandered through wrong handling. There is a series of steps which should be carefully observed in this connection. The carting of the manure out to the field and the ploughing of it into the soil, etc., are steps which should follow one another in rapid succession. When the manure is distributed in small heaps or spread out on the field and left for any length of time, its important ingredients are liable to washing out, drying up, leaching and oxidization. When in this way the soluble fertilizing substances are washed away, and the nitrogen is completely gone, the farmer then wonders why he does not get the same harvest as with chemical fertilizers, or he is amazed that his manure was "poor", or he arrives at the usually satisfying conclusion that organic fertilizing is not sufficient to meet the food requirements of the plant. But if he has done his best to dissipate the nutritive materials of the soil he need hardly be astonished at the poor and insufficient results. In any case the blame should not be put on the organic fertilizer.

How then does the equation of nutritive materials stand? It may be presumed that one head of cattle weighing 1,000 lbs. will give about 13 tons of fresh manure, or that a cow of 900–1,000 lbs. live weight will give 10 tons of manure and 12–15 tons where there is stable feeding the whole year through and a generous amount of bedding straw mixed into the manure. In the case of a mixed or diversified farm (70–75 per cent acres cultivated fields, 25–30 per cent acres pasture and hayland) with stable feeding for half the year in the temperate zones, we may expect, under normal condi-

MAINTENANCE OF LIVING CONDITION OF THE SOIL

tions, about 6–10 tons of manure from each grown animal, e.g. 10 tons from dairy cattle and 6 tons from beef cattle.

One ton of manure—without the conserving effect of the bio-dynamic treatment—includes the following important nutritive materials, in lbs.[1]

The total of nutritive materials taken out of the earth in a three-year crop succession—140–150 bushels per acre of potatoes, 45 bushels per acre of winter wheat, 55 bushels per acre of summer grain—will show the following tabulation in pounds per acre on the assumption that straw and stalks have gone back into the fertilizer:

	Potash	Phosphoric Acid	Nitrogen
Potatoes:	295	33	73
Winter Wheat:	37	59	147
Oats:	31	55	134
Totals	363	147	354

On the basis of the previous table of the ingredients of rotted stable manure, it is shown to be necessary, where fertilizing occurs

[1]

	Water	Organic Substances	Total Nitrogen	Potash
Horse Manure:	1568	558	13	12
Cow Manure:	1705	440	7	11
Sheep Manure:	1496	660	19	15
Pig Manure:	1593	551	10	13
Mixed Stable-Manure, fresh:	1650	462	10	13
Mixed Stable-Manure rotted:	1694	374[2]	12	15
Liquid Manure:	2160	18	6	12
Chicken Manure:	1232	561	34	19

	Lime	Phosphoric Acid	Silicic Acid	Magnesia
Horse Manure:	5	6	39	3
Cow Manure:	10	6	19	2
Sheep Manure	17	5	32	4
Pig Manure:	2	4	24	3
Mixed Stable-Manure, fresh:	11	4	24	3
Mixed Stable-Manure, rotted:	15	5	37	4
Liquid Manure:	0.7	0.2	0.4	0.9
Chicken Manure:	53	34	77	16

[2] To estimate weight of heaps: 1 cubic yard of fresh strawy manure weighs about 1,100 lbs. 1 cubic yard of well-rotted, not too wet manure weighs 1,300

once in three years, that approximately 12 tons of manure per acre be used. This is true provided that we base our calculations only on the bare figures of quantities extracted and substituted, and that we ignore the question of soil quality and its reserves of the necessary ingredients as well as losses through seepage.

Such a manuring per acre is, as a matter of fact, usually given on mixed farms. This total should suffice if we take into consideration the fact that certain losses are made up by means of the previously described, careful treatment of the manure. Translated into terms of cows, with six months' stabling, this means two head of mature cattle per acre of cultivated field, which is the proportion usually attained on small farms.

Experience has shown that on the large, grain-growing farm the proportion is generally not up to this standard. The situation can be improved by crop rotations that are not exhausting to the land, and by the use of catch crops. The planting of legumes is particularly valuable for meeting the nitrogen requirements, still without counting the "reserves" in the soil.

This rough calculation throws light on an important problem, as to which farms are in a biological state of health—that is to say, which farms are healthy as a result of the equilibrium of their cultivated and grass areas, and their crop rotations, and which farms, viewed from the bio-dynamic standpoint, are going in an unhealthy direction due to one-sided practices and thus are farms to which "correctives" can bring no lasting cure. The author has actually seen farms which, although having only one head of cattle per seven to ten acres, were able to keep going. But in such cases all the finesse and skill of experienced farmers are required to preserve the farm's "organic basis".

We can clearly deduce, from the foregoing, the amounts of manure necessary for the adequate fertilization of a piece of land. If care is taken that the full fertilizer values are retained in the manure, no one who follows this recipe need fear robbing the soil or risking poor harvests. It is the chief goal of the bio-dynamic method of agriculture to educate the farmer to this high standard of manure utilization.

---

to 1,400 lbs. Recently, in practice, we have been able to calculate: one adult dairy cow produces 10 tons of manure per year. One adult beef animal produces 6 tons of manure which could be collected. The manure equivalent for various animals is: the manure of one dairy cow equals that of $1\frac{1}{2}$ bred heifers, or 2 yearling heifers, or 2 horses, or 10 pigs, or 20 sheep, or 200 hens or 1,000 rabbits. These amounts would be sufficient for one acre of land.

MAINTENANCE OF LIVING CONDITION OF THE SOIL

Those who read this exposition of the bio-dynamic method may think that only a farmer with the necessary amount of cattle is in a position to start using the bio-dynamic method of agriculture. Our reply to this would be that, in fact, it is true that no farm can have real value unless it is based on a *proper proportion* of pasture, hay land and cattle to tilled fields. The bio-dynamic method of agriculture was not designed for a farm without cattle and for which no organic fertilizer can be purchased. A farm without cattle represents a biological onesidedness and is contrary to nature. Likewise a mere cattle range such as often exists in hilly regions or in places where the soil is too heavy also represents something basically one-sided and needs a regulative, balancing factor. Those who object to the state of balance aimed at in bio-dynamic agriculture have a complete misconception of what a farm should be. Making things grow on a piece of soil is not necessarily farming—it may only mean destroying the earth's fertility.

To return to the care of soil and manure, the bio-dynamic manure—quickly rotted and usually easily spread with a shovel—should be spread out on the ground the moment it is carried to the field and then immediately ploughed under. If it is desirable to cart the manure to the field at a time when other work is slack, perhaps on a rainy day, the entire heap may be taken to the field upon which it is to be used, and there heaped up and covered again. This may be considered as a process similar to the turning of a manure heap. Once the manure is spread out, however, it should be ploughed under within three hours if it is to retain its full value. Neglect of this point can cause great disappointment and loss of crops. Indeed, it is only necessary to walk over the fields nowadays to see what sins are committed against nature and a healthy agriculture.

A slight exception to the above rule of ploughing in manure immediately is when the manure is carted out on to frozen ground and freezes, thus being preserved from disintegration for the length of the frost. This spreading of the manure without immediate ploughing under is also permissible under certain circumstances on pasture and hay lands. It may be necessary in working with very heavy wet soils. In the case of a pasture, a good harrowing should previously have been given so that the field is in condition to take up quickly the thawing manure solution.

When the manure is ploughed under, *care should be taken not to bury it too deeply.* It should continue the process of transforming itself into humus in the ground; and to do so, since this is a biological process, it requires air, life in the soil, and moisture. The most intensive life activity in the soil exists in a surface layer of about from two to six inches deep. The layers below this are relatively lifeless, and should not be disturbed, because there is less aeration in them and therefore considerably less possibility of transforming organic substances. Thus it is of great importance not to plough the organic fertilizer too deeply into the soil; otherwise it will fail to develop its full effect.

Deep ploughing may indeed loosen the ground and serve to mix it and to bring up hitherto unused layers of soil; yet as far as the biological state of the soil is concerned this ploughing offers no advantage, but quite the contrary. By means of deep ploughing the living, upper layer is buried, covered up and isolated and the life processes are brought to a standstill. Manure, for example, which is ploughed under to this lower level, remains—in heavy, cold and wet soils—unaltered for a long time. Months later deep harrowing may bring to the surface the unaltered fragments. Furthermore, the lower, inactive layer of soil which is brought up by deep ploughing will usually need several years to become permeated with soil life, bacteria, etc. In the meantime it is not in a biologically active state. For a long while it still lacks the proper crumbly structure, and the surface tends to be encrusted. Moreover, the pressure exerted by the more compact, *new* upper layer on the crumbly layer that has been ploughed under may lead to the formation of two separate levels, causing water to stagnate and disturbing the circulation. This dammed-up moisture is harmful to germinating seed. The necessity to plough deeper for root crops and potatoes than for grain is obvious. Here we are concerned with fundamental principles of obtaining the maximum value of the manure which must be adjusted in practice to the particular crop and kind of soil.

Let us assume that up to this stage everything has gone well: careful piling up of the organic fertilizer, covering, inserting the preparations, carting out to the field, quick ploughing under to the correct depth. There are still, however, a number of factors remaining, which will determine whether organic fertilizer has been more or less properly utilized. Air and moisture are not the only requirements. We have just pointed to the repressive influence of stagnant moisture. Such a condition can be partially remedied by proper drainage, an absolute prerequisite for healthy biological conditions. Most important, however, is the maintenance of

MAINTENANCE OF LIVING CONDITION OF THE SOIL

capillary action in the soil. Every measure which disturbs this acts as a hindrance to getting the best out of the manure. The most generous manuring, if accompanied by incorrect cultivation of the soil, results in only a half value. In investigating complaints about insufficient yields, I have often been compelled to point to mistakes in cultivation, rather than to poor or faulty manuring. Bio-dynamic farming cannot be carried on without correct cultivation. In cases where this is neglected the farmer should not blame the bio-dynamic method but rather his own mistakes.

The causes of mistakes in cultivation are often hard to bring to light. Although often neglected, the most important and indeed well-known points to bear in mind are: care in the ploughing and harrowing of a wet soil, so as to avoid bringing up clods and the smearing and closing of the surface; and the use of the roller at the correct time, when the ground is becoming dry, in order to retain capillarity and prevent surface incrustation. A soil that is alive, and kept permeated with organic processes, inclines of its own accord towards a crumbly structure. Anyone who maintains a farm or garden of this sort has, by his correct cultivation at the proper time, become a friend to the bio-dynamic method of agriculture in its relation to the improvement of the soil. Often it means having both courage and caution to wait a few days more before working the still too wet heavy soil. The bio-dynamic point of view teaches this caution. A complete and quick change of all plans may be required in order to roll the soil at the necessary time before it dries out and the surface closes in a crust. With care, the experienced farmer can always prevent incrustation. The much maligned "slow" peasant knows that starting the "right work" at the "proper time" brings the *harvest*. Ploughing and manuring are only the preparation. But taking advantage of the correct state of the soil and weather is what generally determines the harvest. If our young budding agriculturists were only given more *practical* training in this matter, a great number of merely passable areas might be turned into first-class farms. But it can all too often be seen how countless valuable factors in the soil are lost through lack of knowledge and care. This is above all true in "young" newly settled lands, where there is no old peasantry wise in the ways of the soil. Half of the dreaded condition of erosion in subtropical and tropical regions is due to the lack of knowledge and care.

A comparison has been made between humus-conserving and humus-consuming crop rotations; and also between humus-conserving and humus-consuming weather conditions of various

SOIL FERTILITY, RENEWAL AND PRESERVATION

years.[1] A crop rotation overweighted on the side of grain, for instance, consumes nitrogen from the soil, while legumes conserve it and, as is well known, also return nitrogen to the soil. They also act on the weathering of the soil, and so set free new reserves. A study of the comparative action of various cultivated plants in opening up the soil is very enlightening. The *Handbuch für Bodenkunde* of E. Blank, Vol. 2, p. 260, in dealing with "biological weathering by living organisms", gives some important figures. For example, it shows that various plants take up entirely different amounts of mineral substances from one and the same soil.

The amounts—in grammes—of mineral substances in a specific rock material brought into solution, and taken up by various plants, are as follows:

	From Variegated Sandstone	From Basalt
3 Lupin Plants	0.60	0.75
3 Pea Plants	0.48	0.71
20 Asparagus Plants	0.26	0.36
8 Wheat Plants	0.027	0.19
8 Rye Plants	0.013	0.13

The legumes have thus approximately a seven to sixty fold stronger effect in breaking down the soil than the grain. This fact, with the recognized value of proper crop rotation and cultivation, as well as the fixation of nitrogen, points to the special benefits of the leguminosae, the use of which at certain times should not be omitted from any crop rotation. Only by the use of legumes can a significant proportion of the natural reserves be brought into use for the fertilizing of the ground. This, together with the treatment of manure and compost, constitutes one of the most important aspects of the bio-dynamic method of agriculture. But too frequent planting of legumes brings about an excess of nitrogen, which in turn hinders the fruiting of plants and causes them to tend toward leafiness. Therefore the legumes, i.e. peas, beans, clover, lupins, alfalfa, serradilla, etc., must always form part of a healthy crop rotation as a *curative factor*; but balance and moderation must be maintained even with legumes.

[1] Wet and cool years are more humus-conserving than hot and dry years.

CHAPTER EIGHT

*How to convert an Ordinary Farm
into a Bio-dynamic Farm*

It is difficult to discuss the general principles of converting a farm to the bio-dynamic system without going into considerable detail. In regard to the important question of planting, we see that every region, soil-type, climate, and set of marketing conditions requires its own special crop rotation. In every case, however, a legume should be grown among the crops every four years. This is often done by sowing clover *with* the summer grain, or by planting beans and peas *instead* of summer grain. Crop rotation extending over a cycle of five to seven years—often the rule in the northern countries—has a more beneficial effect than the "intensive" three-year rotation.

The following is an example of a somewhat too intensive programme for the biological farm: first year, hoed crops (potatoes, beet and turnips); second year, winter grain; third year, summer grain. An improvement in this rotation would be to sow clover among the third-year grains which would then remain over into the fourth year. In heavy, wet soil, near the sea or in a damp, hilly climate, the hoed crop during a moist autumn may ripen too late to prepare the land in time for the planting of winter grain. Then the rotation should be as follows: 1. hoed crops; 2. summer grain; 3. legumes; 4. winter grain. The effect of the stable manure used on the hoed crop remains in the soil for the following summer grain; this is supplemented in the third year by planting legumes; this produces good winter grain with less danger of lodging from moisture. Other variations are also possible.

A further improvement will result from growing grain with catch crops, and green manuring, or by planting legumes after grain. These legumes are especially important for light and sandy soil. Just as in forestry all monoculture has proved itself to be

SOIL FERTILITY, RENEWAL AND PRESERVATION

biologically unsatisfactory, so also has the growing of one-sided field crops likewise been proved unsatisfactory. By the use of the grain plus something else, oats with vetch or broad beans, or something similar, the beneficial effect of the simultaneously growing legumes balances the one-sided effect of the grain. An intercrop of serradilla on a sandy soil is to be recommended. Green manuring with vetch can be profitably made use of in a heavy soil.

Many will say: "Oh yes, we know all this without any bio-dynamic method of agriculture"; but, it may be asked, *who really acts in accordance with this knowledge?* The point is that these rules should be actually *followed*. We have made it fundamental in our programme for the bio-dynamic method of agriculture energetically to put into practice these rules of procedure for the improvement of soil fertility, which while familiar in themselves are generally neglected.

We must also point out one more problem. In a climate of moist, rainy summers, the growing of grain mixed with other crops involves difficulties. The grain is ripe; the catch crop—for example vetch—is still green. When the field is cut, the grain dries more quickly than the legumes which, still moist, have a bad effect on the drying process. Feeding the grain crop green as forage or dry as hay where there is a heavy soil—as for example in Holland—is too expensive as an alternative procedure. For each particular case special consideration is necessary. For example, the farther we go in Europe to the drier east, or wherever in other regions of the Earth we have dry and sandy areas, the more is this growing of catch crops worth practising. Even if the catch crop is only to be used as fodder along with the grain straw, this sort of planting is advantageous since it provides for protein requirements with home-grown feed.

In regard to green manuring we are handicapped in moist climates adjacent to the Atlantic by having very little frost in winter. The simplest green-manuring procedure is to plough the plant mass under after the frost has worked into it. Ploughing the plants into the soil while still green before frost, however, means that the unrotted plant mass takes too much oxidizing force out of the soil and thus diminishes the possibilities of bacterial life. This danger is greater with heavy firm soil than with loose airy soil. Following the bio-dynamic principle it is also better to put the green manure in the compost pile in order to produce a neutral humus. For this it is necessary to gather the material cut green, or even after it has been frozen, and compost it in layers, according to the approved

HOW TO CONVERT A FARM INTO A BIO-DYNAMIC FARM

method. In this way the soil will get humus which it can take up rapidly. This green fertilizer compost is, under favourable conditions, ready for use by spring. The roots are ploughed under. In moist climates, however, we have wet soil, which makes autumn cutting and the gathering of green material difficult. So, in this case, the entire plant mass must be turned under evenly by a shallow ploughing, to be followed later by the deeper winter furrow. If there is a good, crumbly soil structure, and preparation 500 is applied at the time of this ploughing under, this procedure may prove satisfactory. On a muddy, firm soil, however, it is inadvisable.

The planting of legumes following grain is very useful in intensive farming. Beans, vetch, lupins, etc., follow a grain harvest. This brings nitrogen into the soil, and the ground remains covered for the rest of the summer, and is protected against loss of moisture. The resulting harvest is a high protein fodder relished by the animals. The important thing is that immediately after the grain harvest the stubble be peeled off and the ground harrowed, sown and rolled. Only in this way can the beneficial, crumbly structure of the soil be maintained. Every day and even every hour lost in putting in the crop that is to follow—after clearing the field—means loss of water and of soil structure. An experienced farmer knows that *within three hours after mowing the grain, the vetch must already* have been sown. The rows where the grain sheaves are set up should be planted separately later. Such procedure gives an idea of what is meant by quick cultivation.

The questions of crop rotation, and of the possibility of using catch crops or after-crops, constitute important problems in converting a farm to the bio-dynamic method of agriculture. How is this conversion best carried out? First, we should not go about farming at random but make a clear plan. We have already pointed out that it takes several years to rejuvenate a soil which has had its organic structure disturbed, and that the bio-dynamic effect can only be fully developed after this length of time—that is to say when, in this process of crop rotation, the manure is applied for the second time on the same parcel of land. So the plan for converting the farm will extend over at least two complete crop rotations.

The first step in this plan is to arrange a manuring scheme. To the extent that the organic fertilizer is available in sufficient quantity, the conversion can be carried out. *The first practical measure then is the careful treatment of the manure.* If the farmer has recognized the value of this work, then he will be ready for further suggestions. *The second practical measure* is the *careful gathering and handling of all organic masses* which can be turned into compost. The treatment of the fields with preparations 500 and 501 begins in the first year of the conversion in those parcels where legumes are growing, and where there is every prospect of manure or compost being used promptly. If this prepared biological fertilizer is then applied, the first stage of the conversion is completed.

Every farm, which has both a regular crop rotation and a permanent number of cattle, can be converted to the bio-dynamic method in this way. Farms without cattle cannot be developed under the bio-dynamic system. In gardening, where there can be no cattle, the purchase of stable manure or other organic matter is the only way.

Where organic fertilizers are available, but in insufficient quantities, they can be supplemented by buying horn and hoof meal (for nitrogen), bone meal (for phosphoric acid), etc. This is no more expensive than mineral fertilizer, and when the material is composted first this helps soil improvement.

Conditions of soil, local demand and available markets determine crop rotation. No one can be advised to grow peas when they cannot be sold. But the farmer can be advised to grow a fodder legume which he can use as feed.

At the second stage of the conversion, the farmer should increase the number of his cattle, and thus his manure production. He should improve the fodder and gradually change over as far as possible from the intensive feeding of concentrates to home-grown feeds. These latter are healthier for the cattle, and improve the humus of the cultivated fields through their manure. The increased milk yield obtained by feeding concentrates is generally not economically sound, when one reckons on the other side of the ledger the expenses of replacing cattle lost by sickness and by abortions. What is needed is milk production commensurate with a healthy condition of the particular type of cows kept. In calculating the economic value of a cow, both its milk production and its calf production must be considered. An average cow with many calves is more economic than a cow producing much milk but few calves. She is also more valuable because the breed of her calves is improved. That is to say, the cow that has not been exhausted by overmilking transmits more life energy and organic reserve strength to her offspring.

By calculating the value of the manure at its market price, the

HOW TO CONVERT A FARM INTO A BIO-DYNAMIC FARM

surprising discovery is made that the normal combination of pasture and arable land is quite economic in spite of a lower wheat production for market purposes—and this even in a region not preponderantly devoted to dairy farming. In any case, the cattle must be brought up to a *high*, but not to a *forced* standard by means of careful breeding and selection.

The working out of the most practical ratio between meadow and arable land, and the calculation of the most advantageous number of cattle for the amount of manure needed, should be done for each farm individually. To attain the best average yield requires a number of years. It also depends on the capital reserves of the individual farmer, which should be called upon as little as possible. Where ample financial reserves are available, the same goal can be reached in three or four years which might take another farmer eight to ten years of slow building up. If carefully carried out, the conversion will not fail. The high premium for the knowledge that makes this assurance possible has already been paid for by the pioneers of the bio-dynamic movement. To-day their experience is available to everyone.

During the period required for achieving the proper proportion of pasture and arable land, the proper proportional increase of quality in the cattle can also be achieved by intelligent breeding. In this connection an important fact must be stated; high-bred cattle cannot be improved, having already reached the apex of improvement. *If they change at all, it is downward*—they degenerate.

For breeding, strong, average cattle, capable of further development, are the most desirable and the best. We have observed that local types, developed under *traditional* methods, offer the best possibilities for further improvement. The importation of breeding cows ought to be limited to the absolutely necessary minimum. Improvement of the breed can be attained by buying a *bull* from outside. Every *bought cow* brought in from outside can bring in diseases. A stable from which contagious abortion has been eradicated may none the less suffer from re-infection brought in by a bought cow. The author had the best breeding results when he introduced young bulls, not yet a year old, raised on soils that were a little poorer than his own and from a climate that was a little more severe. If the points and characteristics of the animal were good, then it had the possibility, through the bettering of its environment, of further improvement. The observant farmer will soon see that his own breeding, when properly carried out, always

SOIL FERTILITY, RENEWAL AND PRESERVATION

gives the best results (cf. Chapter XIV—reports from various farms).

To sum up the steps of conversion:
1. Consideration of what is *desired* and what can be done.
2. Proper care of the manure.
3. Setting up healthy crop rotation and improving the methods.
4. Improving the quality of the manure.
5. Improving the feeding with home-grown feed.
6. Improving the herd as a whole.

All further procedures of the bio-dynamic method depend upon the judgment of the individual farmer. The steps listed under numbers 1 to 5 above are absolutely necessary to obtain the best results. Maximum results are obtained by means of intensive work by the farmer and, under certain circumstances, by the employment of some additional capital. The question of seed is of great importance. We have already remarked that bio-dynamically raised seed has proved its resistance to plant diseases. Hence, the bio-dynamic farmer must gradually raise such seed himself on his own farm. For example, it is always practicable for a not-too-small gardening establishment to grow a part of its own vegetable seed. For the general purpose of this book it is not necessary to give more details about seed growing. It must be mentioned, however, that seeds for hothouse culture should always be produced on open fields.

Seed culture without strict selection is impossible, even when so-called mechanical cleaning and sorting are practised. If, for example, potato seed is being selected, the stand of plants must be observed throughout their entire period of growth. The strongest and finest blossoming plants should be marked. Of these only those that are surrounded by healthy plants should be used. Finally, those that eventually are best are dug out before the harvest of the others; this makes extra work, but repays the effort.

Another question is that of putting up protective hedges as windbreaks. This is important for pastures, especially for young cattle. And we have observed further that the digestive activity of cattle is helpfully stimulated by their eating leaves from shrubs and trees. A little "nibbling" on the leaves of the hazel increases the butter-fat content of the milk.

There is one question that comes up again and again in connection with the conversion of a place—namely, that all these things require more labour and this adds greatly to labour costs. Thus, it is said, conversion is expensive and will not pay economi-

HOW TO CONVERT A FARM INTO A BIO-DYNAMIC FARM

cally. As a matter of fact, superficially considered, extra labour is needed, for example, to gather the compost, to turn the manure or compost piles, and, in the long run, to make changes in the cultivating methods and the crop rotation. The growing of mixed crops means extra work, etc. But these expenditures should be compared with the amount of time we consume in the purchase, hauling and spreading of mineral fertilizers, in the spraying of copper sulphate or other similar materials, in the working of a dead, encrusted soil, in the purchase and hauling of concentrate feeds, and even in the constant trading in cows. The latter may be indeed a pleasant occupation generally carried out over a cup of coffee (or a glass of beer!) but it bears witness to the fact that something is wrong in the stable. When we count all the time expended in this way, then the whole matter takes on a different complexion.

In this connection, the author is able to cite the testimony of practical men actually engaged in farming. One farmer reports that formerly, on one acre, he used 19 horse-hours of work and 27 man-hours of labour per annum. The conversion required 21 horse-hours and 30 man-hours for the same amount of ground on account of the increased work in connection with manure and compost handling, hauling, setting up, covering, turning. The horse-hours appear high, because this farmer had to travel more than a mile to get to his fields. To balance this, he puts on the other side of the ledger the higher fertilizing value of his manure. When the ripe manure is spread we discover the first reduction of the amount of labour required. It is so well rotted that it can be spread out directly and rapidly with a shovel. The manure is loaded from the heap directly on to the wagon, spread broadcast from it, and ploughed in. Formerly there was an additional step at this point. The manure was taken from the big heap and set up in little heaps on the field, and these were then spread out. This involved additional losses of nutritive materials, especially since several days often passed before the manure was ploughed under. It is advantageous to the farmer to plough under the manure *immediately after spreading*. Controlled experiments in Holland have shown that if the yield of immediately-ploughed-in manure is reckoned at 100 per cent, then the yield from manure that has lain on the field three days is only 86 per cent, and the yield from manure that has lain on the ground for some weeks is correspondingly less. The mechanical "manure spreader" while eliminating the aforementioned intervening stage of the little heaps gives an

SOIL FERTILITY, RENEWAL AND PRESERVATION

almost corresponding loss of fertilizer values, because the plants are unable fully to utilize the *raw, unrotted manure* which is generally spread out by this method for ploughing under. Recently, however, the manure spreader has become a very practical instrument in connection with well-rotted manure if followed immediately by the plough or disc-harrow.

We have already spoken of the fact that the bio-dynamic procedure helps to bring lightness to the soil and to develop a crumbly structure. The farmer whose experience is cited above has also had this result. It became plainly evident in the field work, harrowing, hoeing, etc. The hand hoeing of potatoes on this farmer's place formerly took nine hours per acre; now it takes seven. He reports further that in the crumbly soil the weed seeds can be eradicated more speedily, generally with just the cultivator. The time necessary for stirring and spraying preparations 500 and 501 is balanced by the need in the current method for hauling and spreading mineral fertilizer at the busiest period of the year. Another farmer operating a large farm reports that it formerly required 4 man-days of labour per acre for hoeing sugar-beet fields, while now, with the looser soil, the same work takes 2.9 man-days. He confirms the figures on potato hoeing as seven hours now as against a former nine hours. On a farm of 437 acres he has been able to do with one less team. Thus he has reduced not only the hours but the fodder his horses require.

On a larger farm it is essential to make one or more persons individually responsible for carrying out the bio-dynamic measures. Only in this way can we be sure that everything will be done correctly and at the proper time. Such workers should have special training on bio-dynamic model farms. We should like to call attention here to a social consequence of this new method: it demands a greater interest in the phenomena of nature from the farmer and thus raises him to a higher conception of his calling. Then again, because of the intensification of work, it provides the possibility of resettling more people on the land. The author has had the opportunity of observing a number of "homesteading" experiments. It seems to him that most of them have gone to pieces because the people concerned were not well enough trained in their calling, or had not enough interest in it. A certain amount of exact knowledge is essential. The lack of interest shown by people who have been industrialized cannot be remedied by just settling them on the land and teaching them the laws of nutrition and promising high crop yields.

HOW TO CONVERT A FARM INTO A BIO-DYNAMIC FARM

The right basis has been created only when the homesteader has an inner relationship to his work, that is, when he learns to survey and comprehend the sum total of the life processes of an agricultural organism. He will then also love it as one can only love something that is alive. With refined and wide-awake senses, he will see every individual reaction of the soil, the plants and the animals in that large interconnection which alone signifies health and growth and represents his own future.

We have had the opportunity of observing, in a number of homestead projects in Palestine, how *enthusiasm* and *goodwill* by themselves are as little guarantees of success as the presence of mere *bodily* strength and energy. This we have observed in most cases of rural resettlement. In one case there was abundant enthusiasm but no working ability; in the other, working ability but no enthusiasm nor will to work. A homesteading experiment on a world-historic scale, which was only partially successful, was the settling of the Western sections of North America in the course of the last century. Conditions in those regions to-day offer a good example of what the disregard of the laws of "biological totality" in farms can produce. On the other hand, in sections of America where a good peasant population such as the Pennsylvania Dutch settled, there are still healthy conditions prevailing to-day.

In addition to the plan for converting the fields, one must also arrange a fodder plan. Since our goal is to manage as far as possible with home-grown fodder, its planting must be advanced in the proper ratio. This is done on the basis of our knowledge of the "soil-conserving" crop rotation; above all, the correctly timed plantings of clover, lucerne and mixtures of legumes with summer grain. While we want to suggest several possible programmes, we should like to remind the reader that this theme cannot be treated exhaustively here. The ideal feeding is pasture in the summer and clover and grass-hay in the winter. Small peasant-type farms—above all, those in hilly regions—often come closest to this, with clover and grass pasture, beginning in the early summer, turnips in the autumn, hay and straw in the winter. One sugar-beet farm feeds a considerable amount of clover, straw and legume hays (mechanically dried). Another plan is: May to September, pasture and straw; September to October, green corn (maize); October to November, fodder cabbage (planted after grain harvests in a sandy soil), some potatoes, straw; December to May, hay, mixed straw, fodder beet and the like, and legume hay. The programme of a grain

SOIL FERTILITY, RENEWAL AND PRESERVATION

farm: May to September, red clover, lucerne, straw (when cattle are stabled); October to November, turnip tops, straw, hay, clover hay; winter, mangolds, chopped turnips, a little distiller's mash, hay, straw, bran. These examples could be multiplied extensively. They vary according to soil and climate.

A further important point regarding the conversion of a farm has to do with the value of ploughed-up meadows and pastures. If the soil may be said to have already "rested" after a *year* of legumes, how much more truly this may be said of it when it has lain as pasture or meadow for a longer period. In such a case the result is a biologically quickened soil which, for the first one or two years after ploughing up, needs no fertilizer and gives very good yields. If good yields do not result, we find the cause in a common mistake. A meadow has been ploughed up, and the grain or potatoes planted in it have produced a poor crop. Why? The answer is that the pasture itself had become poor and thin *before* it was ploughed up. It had perhaps been poorly cared for, without compost, or was wrongly pastured or cut. A meadow also needs care; it needs compost now and again. We have found that compost has a markedly better effect than manure on pasture land. If no compost is available, then at least the manure ought to be given in a composted form. The more earthy it is the more easily is it taken up by the grassy soil. There is considerable loss if the manure —especially manure full of straw—remains *lying too long on the grass* and is dried, leached or washed out.

The surface of a pasture or meadow, as well as a ploughed field, must be able to "breathe". It needs aeration as much as does the soil around fruit trees or for that matter all soil. A matting down, a closing up of the surface means a retrogression of the good grasses. The sour ones stay, and we can look for an increasing acid reaction of the soil, a disappearance of the clover, and, in moist climates, an increasing mossiness of the ground. Hence the pasture and the permanent hayfield must be harrowed from time to time in the autumn or early spring. If the soil has a tendency to "mat" then it must be harrowed once, deep enough to cause the soil to become visible. Where the grass is growing well and there is enough moisture, it recovers quickly. Naturally the grass must not be torn out by the roots. For this purpose special pasture harrows are used. If earthy compost or manure is put on the freshly harrowed grass, or is even harrowed in, it is taken up immediately by the soil. Correct grazing or mowing is a part of the necessary care of a meadow. A freshly sown meadow needs a definite amount of time

HOW TO CONVERT A FARM INTO A BIO-DYNAMIC FARM

to grow so that all the grasses take a good hold. A profitable pasture or meadow should always be sown with a mixture of various grasses and clovers. For particular conditions of soil and climate a variety of six or seven grasses should be chosen, including high-growing and low-growing types as well as four or five sorts of clover. If this mixture grows, in the first year the creeping grasses will grow faster than the long stalked ones.[1] After some years the latter will predominate. In time a balance is struck. The essential point is that the important low-growing grasses and the clover be not eradicated at the start. This happens if the newly sown pasture is grazed early the same year when the ground is still wet. It should be mowed early in the first summer, and grazed later. In warm or coastal regions where it does not freeze in winter, but is moist, and where the cattle remain outside in the winter months, there is danger that the wet pasture may be too much tramped down. In that case the pasture should be divided into sections and the cattle moved from one to another thus giving the various parts a pause for rest and recovery.

The important thing is always to keep a corresponding balance between high and low growing grasses. This varies from one soil to another, and can best be learned by inquiry at the nearest

[1] Examples of such mixtures:

For a heavy moist soil—	lbs. per acre
Lolium perenne, English rye grass	5
Lolium italicum, Italian rye grass	6
Festuca pratensis, meadow fescue	4
Poa pratensis, smooth meadow grass	2
Festuca rubra, red fescue	2
Phleum pratense, timothy	3
Trifolium repens, white clover	2
Trifolium hybridum, alsike clover	3.5
Trifolium pratense, broad red clover	3

Another possible mixture for clay soil is:	
Festuca pratensis, meadow fescue	3.5
Lolium perenne, English rye grass	6
Daktylis glomerata, cocksfoot	3
Poa pratensis, smooth meadow grass	13
Cynosurus cristatus, crested dogstail	3
Phleum pratense, timothy	1.5
Trifolium repens, white clover	1.2
Trifolium pratense, broad red clover	3
Medicago lupulina, trefoil or nonsuch	3

Shortages and high prices force one at present to use only 20-24 lbs. of these mixtures per acre, which will do on well-prepared soil.

agricultural experimental station or by reference to the appropriate authority.[1]

These figures are correct when the plants are in blossom at the time of cutting. Farm tests show further that, *before* the conversion, 100 lbs. of starch are transformed into 169 lbs. of milk with 5.21 per cent butter fat, and *after* the conversion to the bio-dynamic system 100 lbs. of starch are transformed into 215 lbs. of milk with 6.73 per cent butter fat.

If we are not grazing but mowing, then the first cutting should not be made too late, as is often the case from the mistaken idea that we want to "*have a lot*". Grasses and clovers have their highest content of feed and nourishment values just before and when in blossom. That is the time when they must be cut. Later they give more "straw" but less nourishment. And if they are cut later the possible second or third cutting is poorer and also the balance of the grass mixture changes to the detriment of the lower-growing types. Properly timed cutting, and afterwards the grazing down of hayfields, are of great importance; fertilizing is equally important. Fresh manure and fresh liquid manure are absolutely harmful and have the same bad effect. The clover gradually disappears and the meadow becomes sour. Only well-rotted, earthy compost is helpful; its preparation has been described in Chapter VI. Only this will keep the clover growing. The disappearance of the clover must be regarded as a danger signal. If more and more weeds appear and if the grassed areas, instead of being green at blossoming time, are white or yellow with weeds—this is to be seen rapidly increasing on the Swiss pastures—it means that the soil is sending out an S O.S. It is then high time to *plough up*, manure and aerate the soil. Just compost on the grass will now no longer suffice.

[1] For clarification we append a table of fodder values.

Plant	Protein Percentage	*Fresh* Starch in lbs. per 100 lbs. green plant mass	Protein Percentage	*As Hay* Starch in lbs. per 100 lbs. of hay
English rye grass	1.3	10.6	3.3	22.5
Italian rye grass	1.3	11.4	4.9	35.6
Timothy	1.0	14.0	3.2	29.1
Average of all Sweet grasses	1.5	13.7	4.0	30.2
White clover	1.9	8.8	4.9	32.1
Broad red clover	1.7	10.0	5.5	31.9
Alsike clover	1.7	7.9	5.6	29.8

HOW TO CONVERT A FARM INTO A BIO-DYNAMIC FARM

Farmers often plough up a meadow on which no clover is growing, and which is sour, producing only a strawy grass. They plant grain and are disappointed when the "conversion" does not function. Thus we see that meadows and pastures represent a reserve of soil fertility only when they are in good condition. Only then can we avoid fertilizing for one or two years. If the clover has disappeared and the grass is poor, we must manure the ploughed-up meadow at once, for we have poor soil that is in need of improvement. It is important not to wait too long before ploughing. As long as we can keep wild white clover (trifolium repens) growing, all will be well. This clover has special significance because it is drought-resistant and can improve the very worst soils. For drought areas, lespedeza has also shown itself to be very valuable. In this connection we must also mention the value of "short leys" (temporary pastures) in the crop rotation.

Agricultural crop rotation extending over eight-year periods may prove practicable. In that case, a grass and clover mixture should be sown. This remains for four years as a hayfield, which is cut several times each year and pastured in the autumn, and which should be given some compost between the third and fourth year. Then it is ploughed up and the normal crop rotation of field crops begins. We call such a rotation the best and most soil-conserving. And with proper care it also gives a high yield of hay. This system is especially suited, perhaps even essential, to a heavy soil in a moist climate in order to counteract souring and the spread of moss. The author has used it with very good results on his own farm in Holland.

Ploughing up of meadows is best done in the autumn, so that the sods can break up well during the winter. The sowing of summer grain then follows naturally in the second year. If the stand of grain was moderate, a hoed crop may be planted with manure. One may also plant peas and then beans immediately after ploughing up. This is advisable when the preceding grass has been in poor condition, and it is good to use some compost or ripe manure with the beans in the second year.

The cockchafer grubs and similar insects in old pasture and grassland are annoying. If we are transforming a small parcel of land, for example, into a garden, it is best to fence it in and let pigs and chickens run on it for a short time. Their presence will produce a loosening and cleaning of the ground. On larger areas there must be an intensive reworking of the soil; it needs harrowing especially to hinder the development of maggots. Light and air are their enemies. If other measures do not help, *"trap plants"* may be employed by sowing spinach or by scattering pieces of potatoes, which are later gathered up covered with these pests, thus protecting the crop planted in the field. Hence, under some circumstances the planting of peas, even with the large amount of hoeing they entail, is advisable on old pastures.

Preparations 500 and 501 have a part in the bio-dynamic handling of meadows and pastures. Preparation 500 should be used in the autumn and once in the spring; preparation 501, after there is no longer danger of night frosts, should be sprayed on the green plant (cf. Chapter VI), and again a short time after the first and second cuttings. We have noted especially good results in dry periods when we applied preparation 500 immediately after the first cutting, and sprayed 501 a week or two later.

When laying out a garden plot where grass has been growing, it is advisable to peel off the sods and make them into a special compost heap. The subsoil should then be worked in a normal way and later this compost should be returned to it. The conversion of a garden offers no difficulties, if bio-dynamically treated compost and manure are present in sufficient quantities and if the principles of crop rotation are observed. An especially intensive application of bio-dynamic methods can be made in a gardening establishment. To outline this, however, would go beyond the limits of this book; the details must be left to the advice of bio-dynamic information centres.

It is our usual custom to-day to think of the economics of a farm in industrial and commercial terms. But this method of thinking about farm economics assumes that there should be a capital turn-over in a year with at least a 20 per cent profit. Otherwise there is no "urge" to start such a "business". This at present is the attitude of mind of those occupied in modern commercial farming; especially in such commercial enterprises as the monoculture of sugar cane, citrus trees, tobacco, and in dairy farming, etc. These are usually started with a very speculative point of view. This monoculture may succeed if its returns begin within a very few years—4, 5 or 6 years. The longer it takes to reach the productive period the worse the economic value of the enterprise. This is because this one-sidedness never permits the soil condition, the humus structure, the general fertility, to improve.

The production of foodstuffs is one of the most important and vital problems of humanity, and is hardly the most fitting field for speculation. The economic side of food production should be

HOW TO CONVERT A FARM INTO A BIO-DYNAMIC FARM

calculated on the basis of a long-term rhythm, like the rhythmic advance of the evolution of the human race itself.

Experience shows that the proper basis for an economical, self-supporting farm—i.e. the basis for the capital *turn-over*—rests on the crop "turn-over", the rotation of crops. This means that we cannot expect the return of the capital expended in one year. Its return is proportional to the crop rotation. If we have a three-year crop rotation the capital turnover takes three years to accomplish. This is in harmony with a natural biological rhythm. If we have a five-year crop rotation, the capital turn-over requires five years. This gives farming a healthy economical basis. The farm then is really a "hedge" against a "boom" in the one direction or a "depression" in the other. It has been discovered, as a corollary, that the shorter the term of crop rotation the more intense the work; a three-year turn-over of capital requires more implements and a greater intensity of labour on the farm than a five-year turn-over. For example, we must put in relatively more money for implements, for labour spent on the soil. The longer the period of crop rotation the less intensively do we have to work, and therefore the lower are our costs. There is less rush, less effort, less strain. One main crop lost in a three-year rotation period means the loss of one-third of the arable land (production value). In the five-year rotation plan, the loss is only one-fifth and the economic condition is more "elastic" and "resilient" in its resistance to outer influences. At the same time we observe that our efforts are more "biological"—for we are saving the humus and the soil fertility. We thus learn by practice that the biologically most balanced farm is at the same time the most self-supporting farm economically. The profit is small. Under present conditions a profit of from 2 per cent to 3 per cent is relatively high. But it is fairly well guaranteed and the capital is safe—provided, of course, the treatment of the soil saves the "soil capital"—the *humus*.

Recently the value of a diversified crop has been recognized in the more intelligent official circles in Europe, where it has been discovered that only on soils where the ideas presented in this book have been properly applied are really healthy farms to be found. A practical state of society (in the past) introduced the idea of the "family farm". A farmer and his family did the work and kept the farm intact and fruitful for generations. The idea, both of the "family farmer" and of "diversified farming", seems to the author to be the only basis for a really healthy future farms life and healthy social conditions. *But this is true only*

SOIL FERTILITY, RENEWAL AND PRESERVATION

when the full effect of the humus-saving bio-dynamic system has been attained.

"But does bio-dynamic farming pay?" is the question generally asked after there is acknowledged agreement on the value and importance of bio-dynamic principles for the conservation of the soil's fertility.

After a careful study of farming economics on many farms and in many countries, in times of farming booms as well as depressions, the only true answer I have found to this question is another one, "Does any farming pay under your particular circumstances?" If it does, then bio-dynamic farming will pay too. If not, then we must search for the reason why and try to eliminate the trouble if possible. Only then, after the bottleneck has been cleared, can we decide whether or not bio-dynamic farming will pay.

The expenses of purely bio-dynamic practices in proportion to the general expenses on a farm are so small as to be negligible. The additional income, however, may be influenced by so many factors outside our control—weather, markets, the labour situation, etc.—that it is difficult to say how much should be credited to the bio-dynamic method in particular. One should certainly not lay at the door of bio-dynamics the burdens of labour, market, investment or mortgage situations, inadequacy of the soil—factors which would exist on the same farm in any case.

A direct practical example will best prove our statement. Wilbur Jones has a 100-acre farm with 20 head of dairy cows, 40 acres of pasture and 60 acres tilled with a five-year crop rotation which allows him to grow a major part of his feed, fill his silo, raise a few heifers, pigs, etc. Milk from his cows brings an income of £1,600 annually. In addition, he sells a few calves, heifers, pigs, eggs, a few bushels of grain, executes a few local orders for neighbours, etc., so that his total income is £1,875 per year. His expenses are £1,500 a year including everything—feed, petrol, taxes, insurance, mortgage interest and gradual repayment, but no depreciation. In his enthusiasm for bio-dynamics Mr. Jones treats all his fields, pastures and manure and compost heaps with the preparations. He needs 75 portions each of 500 and 501 at 9d. and 6d. each, totalling £4 13s. 9d. He also prepares 200 tons of manure and 100 tons of compost at the rate of one set of 502 to 507 per 15 tons, making 20 sets at 2s. 6d. each. Large orders for preparations all sent at one time being subject to 10 per cent discount, this makes his outlay on the preparations £6 9s. 5d. per year. He has additional labour for stirring and spraying, which according to our

HOW TO CONVERT A FARM INTO A BIO-DYNAMIC FARM

experience amounts to about 68 hours for this amount of 500 and 501 if the spraying is done by hand (in about 48 hours), and which at 1s. 6d. an hour would be £5 2s. We have seen 100 acres sprayed in 8 hours with a spraying machine. There remains the extra handling of the manure, setting up of heaps, covering, preparing, sometimes watering, loading on a spreader and distributing on the field. We have had cases where in practice this extra labour on manure amounted to 9d. per ton or even as low as 6d. With beginners or very slow people it may amount to 1s. 3d. a ton. Reckoning on an average of 9d. per ton (half an hour per ton) 200 tons would amount to £7 10s. Summarizing, we then have the following:

	£	s.	d.
Preparations	6	9	5
Labour, 500, 501	5	2	0
Extra labour, manure	7	10	0
	£19	1	5

Of Jones's total expenditure of £1,500, this is only a small percentage which will certainly not be the decisive factor to tell whether the farm pays or not.

One comparison may be of interest, the expenses for maintenance of the farm—machine repair, paint, building repairs, gates, fences, etc., which may easily run to £40—50 per year. Is it surprising, then, if Mr. Jones spends this amount for maintenance of property, that he should spend £19 1s. 5d. for maintenance of soil fertility by improvement of its humus content? He certainly gains by so doing no matter how his farm economics behave in other directions.

What may he gain by applying the bio-dynamic method? This can be illustrated by various examples; first there is the question of compost. The collecting of refuse of all kinds, provides a material rich in potassium and organic matter, medium in phosphate and fair in nitrogen. The fertilizer value of these ingredients per ton of compost is about 10—15s. provided that the garbage, refuse, weeds, leaves, etc. were immediately composted and not exposed for six months to sunshine, rain and frost before being piled. To collect this material economically requires some experience and skill, but when it is properly done experience has shown that the labour expenses of a completed bio-dynamic compost heap will be about 9s. 6d. per ton. Beginners may take more time. The most

SOIL FERTILITY, RENEWAL AND PRESERVATION

expensive heap known to the writer cost about 12s. per ton. Normally for a 100-acre farm with 100 tons of compost one may reckon on saving £5 5s. in important ingredients which would otherwise be lost and have to be purchased. Good farmyard manure is valued at 15s. per ton in fertilizer value of nitrogen, phosphorus, potassium and other ingredients. Ordinarily one loses 2s. of the 15s. value through washing out, decomposition of nitrogen products, weathering, etc. Of 200 tons per year, one would therefore lose £20. Bio-dynamic treatment avoids most of these losses. Practically one may reckon on 6d. instead of 2s. loss per ton, making £15 worth of nutrients saved on 200 tons of manure. Without bio-dynamic treatment an adequate amount of these nutrients would have to be replaced at a cost of perhaps £16 in addition to the £15 lost, thus making an expense of £31. Even though we reckon on only the first £15, treated manure and compost on Wilbur Jones's farm would represent an increase of £21 5s. in value of organic fertilizer which would otherwise have to be purchased. This sum is to be credited to the original bio-dynamic expenses of £19 1s. 5d. which it will more than cancel, so that one has an actual cash saving as well as an increase in fertility and humus content of the fields.

There are also further gains—an increase of biological activity in the soil, accompanied by an increase of humus-forming bacteria and earthworms. One acre of very fertile soil contains up to two million earthworms producing 40,000 lbs. of humus per year. The nutrients contained in this humus are the equivalent of 10 tons of manure or a minimum of £7 10s. per acre.

Even though one may not appreciate the value of these gains, the following facts are clear. If we continue year by year to lose our 13 per cent of nitrogen and other farm fertilizer valued at £21 5s., in twenty years this will amount to a loss of £425. If we have to abandon the farm on account of its depleted soil and buy a better one, this sum, at a rate of £20 per acre, represents 21 acres or, at £10 per acre, 42 acres. In twenty years we have spoiled at least 21 acres' worth of farmland, or one-fifth of Mr. Jones's farm. Facts of the past make an interesting comparison. Soil scientists give figures to show that it took approximately thirty years to deplete the once fertile American soil and bring on such catastrophes as the dust bowl, floods, and erosion, necessitating the tremendous and important soil-conservation programme of the Agriculture Department at an expense far exceeding the accumulated losses of all the opponents of Mr. Jones's practices in

HOW TO CONVERT A FARM INTO A BIO-DYNAMIC FARM

preserving soil and organic matter. These opponents have depleted their farms in sixty years to the zero value of farmland. They had to spend more in fertilizer and use more government money than their land was worth in order to keep going, or else abandon the farm.

Whether or not biological values show immediately in cash values, economically we cannot lose by following the bio-dynamic line. But then a farmer who is in a tight spot financially—and what farmer is not at some time?—may say, "Even if I have increased my earthworm population at the rate of £7 10s. per acre, I cannot sell them, so what?" While agreeing that advantages may not be fully evident in the first year, or even the second, I would ask if he had not observed certain things. "Did not your treated manure, being so well rotted, spread more easily and in less time than your bulky, strawy material in earlier years, thus almost saving the time spent in piling it up? Due to the increase in soil life, the more crumbly soil structure with less hard crust, did you not gradually save time by easier ploughing and harrowing?" "Yes," my friend K. would interrupt me, "that is true. I have a rather tough piece of land, 36 acres which I needed to disc three times to break the crust and get a fine soil for sowing. I have followed the bio-dynamic instructions for four years and now get the same land ready with two discings. You can work out for yourself what it saves to disc 36 acres once less." Another big farmer who used to work in an old-fashioned way with five teams now manages with only four, saving one man and the feeding and care of two horses after working for six years to improve the organic humus state of his soil. Now Paul Smith interrupts the discussion to say, "That may do for Mr. Jones. He has owned his farm for many years, has always carried on good farming practices and has only just introduced the new methods. In my case, the farm was bought a couple of years ago. I am convinced of the truth of the bio-dynamic method and have done everything exactly as you prescribed, but it has cost me considerably more than you have said."

"Well," I say, "let us look into your case. How was your farm run before you took it over?" It appears that the former owner sold the farm because he could not make a living there any more. Some of the fields had not seen manure for ten years, some had been planted with corn for the last two years, and so on—all the things which go wrong if good farming methods are abandoned. A soil analysis finally revealed that John Smith's soils were almost down to zero in phosphate and nitrogen, and poor in everything

SOIL FERTILITY, RENEWAL AND PRESERVATION

else, including humus. These were the typical mineralized soils which appear towards the end of the thirty-year period mentioned above. "Of course," I say—and I am sorry to state that I have seen many such cases—"you have to rebuild your soil first of all, regaining the fertility your predecessors lost. That is expensive. But you would have these expenses either way, bio-dynamic or not. It is not the bio-dynamic method which made your problem so costly, but the general biological situation which you inherited. You are paying for the mistakes of the past generation of farmers before you can even make a start. That is not your fault, nor is it the fault of the bio-dynamic method, which only tries to stop the disintegration and help to rebuild organic matter. It is a situation which in the U.S.A. is recognized, for instance, by the Soil Conservation Service which supports the farmer with lime and phosphate in order to help him rebuild the soil. This support, however, is not enough because it does not yet deal with the organic and humus situation. Incidentally, it is interesting to note that the government subsidies to the Jones farm (in the U.S.A.) are actually reckoned at about £40, a figure in line with what we have calculated as "maintenance expenses".

"You are", I would say to Paul Smith, "in a situation like someone who has a farm without a barn and stable. He has to get them no matter what method he follows. So you have to build up first what has been lost either through your own or others' sins. That is why we bio-dynamic farmers are so much opposed to all soil-mineralizing practices—because we realize that soon there may be a situation where the individual farmer is powerless. His means will not allow him to restore further what has been lost, whether the fault was his own or that of the economic system of his age. However, if he becomes soil conscious and decides to restore the fertility of his soil by his own sacrifices, with government help or by any possible means, he is co-operating in the important task of rehabilitating this good earth of ours. This is where the bio-dynamic method comes into its own."

Do we know how much the rehabilitation of that which man has already destroyed will cost, or whether it will pay? No, we do not, but we will have to do it anyway because our future existence depends upon it.

CHAPTER NINE

Comments on Forestry

Forestry can teach us much about the various aspects of soil biology and plant associations. The same laws which we see working themselves out over great stretches of time in the forest, are—when compressed into a few months—applicable to field and garden culture. The formation of immature humus is an example. Numerous leaves and evergreen needles fall to the earth and are cut off from access to air. This in time produces a sour humus. Lack of air, with otherwise satisfactory organic (but unrotted) fertilizer material, always produces sour humus. In both cases lime and ferrous salts, for example, are washed out of the upper layer and deposited in the lower. In this way the so-called meadow-ores are formed. They consist in every case in the development of a middle layer which isolates the upper and lower soil levels. They interfere with the ground water circulation and mark the beginning of a "sickening" of the soil that leads to infertility.

If these hard layers are not very deep they can, of course, be cut up with a subsoil plough. Many deep-growing plant roots also penetrate this hard layer. The stinging nettle is such a plant, which also helps to decrease the iron content in the soil by the formation in it of free ferric oxide. In woodland soils, the yellow locust tree (Robina pseudacacia) can be of valuable service in the same respect.

The situation is more serious when the isolating layer lies deeper, as is frequently the case where steppe-land is in the process of formation. Despite our advocacy of an intensive soil culture in farming and gardening, we cannot in general advise the cultivation of the soil of woodland areas because of the expense and meagre results. For experimental purposes the soil may occasionally be torn up with the sub-surface plough in order to bring air to the unmatured humus layer. But here, too, the effort to find a natural solution of the problem ought to be emphasized. In earlier centuries of European forest economy, the peasants used to drive their pigs and cows into the woods. If this is done with discrimination

SOIL FERTILITY, RENEWAL AND PRESERVATION

and only on a reasonably small scale, it has its value. For it provides not only a loosening of the soil but also adds manure. But it is regrettable to have to say that woodland pasturing has been overdone; the pigs injure the roots, the cows eat the seedlings and the shoots of young trees; as a result many woods are ruined. Use of animals as a regulative measure however, tried experimentally and in a small way, is often most interesting. A movable paddock containing relatively few animals, small enough to be set up and moved frequently from place to place, but large enough so that no "intensive" churning up of the soil occurs, can be very well used for loosening the soil, getting rid of weeds, etc.

If the farmer is in a position to give his pigs such an outing, it helps to make them healthier animals. In this connection, one is reminded of the famous, strong black pigs of Monte Cassino near Naples which are pastured in this way. Chickens can also make themselves useful in this fashion. As a flock they spoil the soil, but a few in a large yard can get rid of the mossy growth on turf, help aerate the ground, and in addition eat troublesome larvae and other pests. Chickens, as well as animals used in this way, must have their yard moved about frequently.

An important factor in forestry is the question of monoculture versus a mixed stand of trees. With very few exceptions (for example, the oak) monoculture has shown itself to be harmful. In monoculture the roots of the trees take nutritive substances from the soil in a most one-sided way. The single crop has an unbalancing effect on the condition of the soil's acidity, and the falling needles or leaves produce only an unbalanced humus, If, for example, only beech leaves fall to the ground, they bake together in time into a thick, impermeable layer, without mixing with the soil. The fertilizing effect of tree leaves is frequently not fully realized.[1]

If we keep in mind that the falling leaf masses are rich in organic acids (tannic acid) and also in mineral acids (SO_4) we shall see here again a source of substances which can help in opening up the fertilizing resources of the soil.[2]

---

[1] In terms of pounds per acre in a normal stand it amounts to:
    3,614 lbs. in the case of a beech wood.
    2,232 lbs. in the case of a spruce wood.
    3,261 lbs. in the case of a pine wood.

[2] Two pounds of dry leaf material, for example, contains SO_3 in grammes as follows:

Fern leafage	1.57	White pine needles	0.62
Wood mosses	1.10	Spruce needles	0.47
Beech leaves	0.73	Pine needles	0.35

COMMENTS ON FORESTRY

The last hundred years have definitely proved that monoculture in a forest is harmful. The unbalanced soil reaction has already been referred to. Another point against it is the swift and unhindered spread of pests and plant diseases. Then, too, a one-sided stand of pine in a dry region is especially to be feared in case of forest fires because of its high content of pitch. The mixed stand of trees of various sorts, both evergreens and the deciduous varieties, has proved itself everywhere to be the most stable biologically and the most advantageous.

We have still to consider, in this connection, the mutual influence of plants upon one another. The sum of knowledge of all the positive and negative influences of plants and their roots on one another, and their mutual "natural" compost fertilizing, is still very small to-day. Yet in spite of our limited knowledge we do recognize the vital significance of this interdependence for the health and biological capacities of plants and trees. For example, with the oak any other tree may be grown. The beech also, in a mixed stand, has a beneficial effect. The spruce, on the contrary, is a robber. It suffers no other tree in its vicinity. It spreads itself out and in time chokes all other trees, except when growing in the neighbourhood of the birch, where it is *greatly stimulated*. We can thus study the mutually beneficial and harmful influences of plants in every climatic region. It might be possible in this way to formulate the fundamental principles of a healthy forestry.

Forests of mixed trees produce a humus that has many components and is therefore loose, crumbly and not impermeable. The spread of harmful insects is checked. In short, we have here the most beneficial biological conditions and, since the fertilizing of forests is not practicable, we need only allow the laws of biology to take their course.

There is a decided difference between forest and garden in the matter of fertilizing, because fertilizing with organic manure is, of course, absolutely necessary for agriculture but not for forests. Moreover, in the judgment of all experts whom we have consulted in various countries, mineral fertilizing of forests has proved a failure. For quality, a definite specific speed of tree growth is necessary. An increase of soluble salts in the ground induces the tree to take up more salts. Since a balanced salt concentration is always present in the plant cells, if more salts are taken up by tree or plant, more water is needed. But more water means a *swifter growth* and a *looser, less firm pulp* with larger cells; mineral fertilizing also involves a greater withdrawal of water from dry soils.

SOIL FERTILITY, RENEWAL AND PRESERVATION

An interesting observation may be inserted here. The desire for higher yields is often expressed by the farmer and is frequently the force that drives him to do unwise things. He thinks of intensive fertilizing only as the means of attaining larger results. But it must be remembered that water is one of the most important elements in the plant: from 40 to 80 per cent of its green mass consists of water. Increased yield therefore means, among other things, more water. An intensive fertilizing of plants is therefore, in general, possible only when enough water is available. In Central Europe, the ground water level has been sinking for a number of years; hence the water situation will make an increased yield impossible if the water level is not regulated. The average lowering of the water table in ten years has been about five feet, although regions are known where this sinking has gone as far as fifty feet. Furthermore it has become evident that canalization and the diverting of streams and brooks, and *the building of artificial backwaters and hydraulic power plants, frequently act unfavourably on the ground water table.*

Since this question is already becoming a burning one in a region well supplied with water, how much more important must it be in dry regions with both a scarcity of water and an intensive grain culture. Humus in the soil holds moisture, prevents an early saturation and sets the moisture free again slowly, so that the ground remains damp far into the dry time of the year. Heavy soil does better in this respect than light soil. But when a heavy soil, poor in humus, does dry out, this produces the worst of all conditions. Lumps are formed as hard as cement. The footnote table shows some percentages of water retention for various classes of soil.[1]

The general regulator of the water economy of a region is its stand of forests. Forests are magnets for the clouds. They hold the rain water and represent a natural reservoir. A region poor in forests is poor in water—excluding, of course, the possibility of

---

[1] In the Humphrey Davy tests the soils are powdered finely, dried at a temperature of 212 degrees Fahrenheit, then set out in the air for one hour, after which their absorption of atmospheric moisture is determined:

Infertile, purely mineral earth	3 per cent
Coarse sand	8 per cent
Fine sand	11 per cent
Normal, medium field soil	13 per cent
Fertile alluvial soil	16 per cent
Fertile soil very rich in humus	18 per cent

The same principle applies in the absorption and retention of warmth.

COMMENTS ON FORESTRY

irrigation. Herein lies the tragedy of China and of conditions arising in the United States. (See Chapter II.) *An urgent, intensive afforestation of the Chinese hill chains, and of the plains of the American "dust bowl" and adjoining regions, is the only salvation.*[1] But this afforestation is not something that can be accomplished in the course of a few years. The proper development of a forest demands definite environmental conditions. The previous complete stripping of comparatively large forest areas has already disturbed the biological balance. The humus layer, the moisture of which was formerly conserved because it was well covered and shaded, is suddenly exposed to wind and sun when the land has been stripped of all its trees. The soil life is disturbed, vegetation hostile to the development of woodland spreads over the ground. On sandy stretches this process is connected with the formation of infertile heath land. Stony hillsides remain bare; in southerly regions they are gullied by erosion. Hence the careful forester does not cut brutally into a stand of trees but "picks his way", that is, each year he takes out a tree only here and there so that others may develop further, and get more light, while the ground always remains shaded. Only thus can a stand of trees remain always "young". If we go through a forest managed in this way at felling time, we shall be amazed at the number of trees lying on the ground while "the woods" are still "standing".

Woodland trees are accustomed by nature to growing up in the shade of others. Even the trees that later need more light prefer shade in their earlier stages. The forester must see to it that in the forest the natural spreading of the seeds is helped. Birds are his helpers here. In every case fenced-in, protecting enclosures should be put up, so that the young plants are not devoured by game. Within these enclosures the ground can be somewhat loosened and given a small dose of bio-dynamic compost. Then we shall observe in that spot an especially effective rooting of woodland seeds. Keeping a close watch on these natural processes seems to be the chief task of the forester. He must take part, as a friend and helper, in the course of growth, and in maintaining the natural state of balance.

The use of bio-dynamic compost, in the planting hole, has also proved itself very valuable in nurseries of young trees. Such nur-

[1] The recent World War has drawn heavily on forest resources all over the world. Discussion of reforestation and of conservation of the remaining woods should become an important item of peace and U.N. conferences, that the scars of warfare may be healed.

SOIL FERTILITY, RENEWAL AND PRESERVATION

series should, as far as possible, always be laid out in a clearing between grown trees, that is, within the atmosphere of the woods.

The locust (Robinia pseudacacia) has shown itself valuable as a protective plant for afforestation and inter-plantings. It is a legume, aiding nitrogen fixation. It has deep-growing roots, which give it a connection with the lower soil levels and the possibility of maintaining itself in dry times and in arid regions. It is relatively without any requirements. A good procedure is to set out protective strips of locust, behind which the rest of the afforestation can be carried further. The well-known German surgeon, Geheimrat Professor Bier, has in two decades developed a model forest on dry sand in the eastern part of the Province of Brandenburg. He has done this with the help of protective locust trees and elder hedges and by supporting the natural seeding process by the use of other assisting plants.

It should be an axiom of forestry, that a forest should never be completely cut down. The art of forestry lies in a properly timed felling programme, and the proper spacing of new growth on the land.[1] On hills or levels where there are no woods, it is not possible simply to plant young trees. They will be unable to find any "woods earth" there, and in most cases the trees will die off again after a few years. Here one must study nature's method of afforestation. On a dry plain, certain weeds are carried by the wind to the moistest places, most protected from the wind. The broom, for example, grows on sandy soil. We see here again the important influence of the legumes. Under the protection of broom and with the help of its soil-improving qualities (Cf. Chapter XI), other plants can develop, small shrubs, perhaps even scrub oak and locust, for instance. Here the human being must take a hand in guiding the natural forces. Locust and broom, if held in check, supply the necessary environment for tree growth. If allowed to run wild they act like weeds, choking everything else. It is impossible here to give the names of suitable protective plants for every climate and soil. The intention is only to demonstrate a principle.

A whole series of shrubs can grow under the aforementioned shelter. The hazel and elder in particular give shade for the starting of young trees. Their fallen leaves and root activity produce a humus in which woodland trees like to grow. We must, in this way, slowly build up a forest, starting with a ground cover, follow-

---

[1] High wages and low economy have often been cited as reasons for total cutting, but recently even in money-minded America selective cutting has been acknowledged as sound practice.

ing with larger shrubs as protecting plants, and finally arriving at the planting of the desired trees. Such a programme needs from three to four years in its preparation, but obviously gets more quickly to its goal than does the direct planting of trees which die off after two years. If one is dealing with the afforestation of dry regions, it is wise to begin in that section where the relatively best conditions of rain, moisture, or dew still prevail and from there slowly extend the edge of the woods into the poorer sections. A farsighted grasp of the situation is essential here. One must begin at a point that is still touched by moist winds in order to create a stand of moisture gatherers, and from there move out slowly to the dry sections. In America this would mean an afforestation of the Rocky Mountains at points where forest possibilities still exist, and proceeding outwards from the edges of any still existing wooded areas. And afforestation could be started along the upper reaches of the great streams, etc.

This plan would work much more quickly than the planting of trees on some isolated spot. In America as in Italy, we have witnessed one of the greatest errors that can be made by man—namely, *complete deforestation*, by felling the cover of mountains and hill tops. The water gatherers on the hill tops are thus eliminated and then the hills themselves become eroded and bare. The water sinks to the valley, forming marshes—and thus both hills and valleys become useless. A properly wooded hill top, on the contrary, creates reservoirs of water and fruitful valleys.

A complete cleaning of the fallen leaves out of the woods and forests has always proved harmful. The fertilizer that the woods themselves produce should be left to them. Leaving only the small twigs and a portion of the fallen leaves as a soil cover, however, has been found satisfactory. The point is that the woods should not look as though they had been swept clean with a broom; for all the life in the soil and undergrowth must be protected.

In a wild wood in which everything grows without let or hindrance, the trees mutually choke themselves. Woods, naturally, need a regulating hand, for otherwise with time an unhealthy, unbalanced condition will develop.

CHAPTER TEN

Comments on Market Gardening

The reader may perhaps ask why an outline of the biological laws of the forest should precede the chapter on gardening. There is a very simple reason for this. The basic laws of forestry—for example, the constant ground cover, the mutual help of various plants to one another, protective planting strips, and a compost fertilizing as natural as possible—are laws which apply in a similar way to gardening. If all that can be learned about the biology of the forest, about perennial forest plants, be compressed into a single year, it will be discovered that the knowledge thus gained of forest conditions makes it possible to establish a really intensive garden culture.

For every soil which has a tendency to dry out—be it garden or farm in a temperate or tropical climate—mulching and shading create conditions under which a healthy fermentation can develop, and losses of humus be avoided. For intensive gardening it is especially necessary that we pay attention to these things.

A garden must also be protected against drying winds. The hedging necessary for this keeps the garden warmer, which in turn brings a ripening of the vegetables several days earlier, and this means greater profit. Such an arrangement comprises a protective strip of hedge plants or trees all round the garden. As we have already seen, a hedge six or seven feet high keeps the wind off the ground for a distance of more than 300 feet and raises the soil temperature in spring from one to two degrees centigrade. The south side and the direction of the prevailing wind are especially to be considered. Where there is heavy wind pressure, it may be that only poplar or wild cherry are possible as windbreaks. For wet soils the alder is recommended because of its nitrogen nodules and the draining effect of its roots. The alder is especially suitable as a tree or shrub alongside ditches and canals. Other good shrubs and trees for hedging purposes are the hazel, blackthorn and birch. This is of course only a hint at the possibilities.

COMMENTS ON MARKET GARDENING

In a warm, misty valley, there should be protection against the cold winter winds, but at the same time an air circulation should be provided. This is especially necessary where fruit trees are planted. The fruit tree needs air currents for its ripening process. With the exception of espalier varieties, the fruit tree should, therefore, not be protected by a wall or too thick a windbreak hedge. Lichens and moss are the signs of stagnant air moisture.

On a larger tract of land, protective hedges should be employed to separate individual fields as well as to enclose the entire area. If the main hedge is composed of a line of shrubs or trees, the inner rows can well be fruit trees, or berry bushes. Provision should always be made carefully for alternations in planting rather than for monocultures. The more varied the planting, the better is its biological effect. The same is true here as for the wood. A large area of apple trees only, or currants only, is much more susceptible to the attack of pests than one broken up with varied plantings. It may be objected that this makes harvesting more complicated. This may be so; but is not the spraying of copper sulphate, arsenic and lead preparations also somewhat complicated and troublesome, and detrimental to the health? By means of other preventive measures still to be discussed, this sort of spraying may be avoided. The amount of labour needed for carrying out one method or another is found to be about equal when the work in each case is considered as a whole.

The area planted with vegetables should in its turn be broken up by rows of high and low growing plants. These give shade in times of drought, help bring about healthy soil fermentation and provide a windbreak. To obtain the beneficial effect of legumes, runner beans, staked peas, etc., are useful. Sweet corn can also be used as an inner hedging plant. In the shelter of these rows, all the lines of low-growing plants can be set out. The most intensive planting system of the Chinese, which often employed as many as six varieties of plants side by side, has produced good results for thousands of years. In their case, special ridges are made in the ground. One kind of plant is set on the ridge. This grows swiftly. Another kind of plant comes between the ridges. Shaded and kept moist, this grows unhindered by dryness. When this latter plant get higher, the one on the ridge is ready to harvest, and the plant growing between can then in turn shelter the ridge.

We learn from forestry the value of a mixed culture and the disadvantages of monoculture, and from agriculture we learn the necessity of a healthy crop rotation for gardening. Here, too, there

SOIL FERTILITY, RENEWAL AND PRESERVATION

are plants which do not harm the soil—such, for instance, as all the legumes. There are also plants which exhaust the soil—as is the case with almost all the cabbage family, especially cauliflower, and also celery, cucumbers, and leeks. Among those plants that take little from the soil are carrots, salsify, beet, radishes, small turnips, onions, lettuce varieties. The point here is to begin by working out the most beneficial crop rotations. In place of the manured and hoed crops in farming, we have in the garden early potatoes and cabbages which also need manure, and which are among the heaviest consumers of the substances of the soil. It is especially to be noted that early potatoes or maize can be planted with good results in rows with bean rows between.

After these, as the next stage in our rotation, can be employed crops with less soil requirements, such as field salad (European corn salad), spinach, kale, kohlrabi, cabbage lettuce, endive and brussels sprouts, provided these latter are not planted after cabbage, but after potatoes, for example. The less soil-exhausting plants, such as spinach, early root vegetables, field salad, lettuce and kohlrabi are particularly suitable as preceding crops.

But it has been shown that not only plant rotation but juxtaposition of plants has an important effect. We need only try planting tomatoes and kohlrabi in a mixed culture to be convinced of the bad influence of the kohlrabi on the tomatoes; or we may plant two to three rows of radishes and alongside them two or three rows of garden cress, and another test planting of radishes with chervil growing alongside. Then, if a "control planting" without border plants is made, a very clear difference in the taste of the radishes will be noticed. Radishes from the control bed, without border plants, are relatively tasteless. Radishes with chervil as border plants have an excellent flavour.

If we wish to have tender radishes in summer, they will do particularly well if planted between cos lettuce. This will be discussed from the scientific aspect in Chapter XII. Here we wish only to give what has been observed and tested in practice in many localities. The author has for many years directed a commercial vegetable garden of about 21 acres. For a long time monoculture had been carried on in this establishment. There had been large plantings of only spinach, only beans, only cabbage, etc. Relying on experience, we gave up that method and to-day follow completely, and profitably, the mixed system. Here, too, one must naturally proceed with common sense and not in an abstract fashion. For example spinach, maize field salad or cos lettuce should not be put in alternate rows with other plants, but in beds. Everything that is transplanted does well in single rows. When beds are planted, they should be alternated just as the rows are, that is, always with legume beds between. If desired, carrots and peas can be planted in alternate beds, with radishes or cos lettuce in rows among the carrots. Mutually beneficial when grown alongside one another are leeks and celery; carrots and peas; early potatoes and maize; cucumbers and beans; cucumbers and maize; kohlrabi and beet; onions and beet; early potatoes and beans; tomatoes and parsley. A further study of such relationships will certainly continue to produce much useful data.

Harmful combinations are tomatoes and kohlrabi; tomatoes and fennel; fennel and dwarf beans. All the aromatic herbs and pot herbs are good as border plants. Other beneficial combinations to be noted are: turnips and peas; dwarf beans and celery; cucumbers and peas; cucumbers and dwarf beans. Where heavily consuming plants have been grown with manure, they can well be followed by dwarf beans, but not by peas. It is more difficult for the latter to stand the direct after effect of manure.

An example of good combinations would be: two rows of celery, alternating with two rows of leeks. Now and again, using a wider space than usual between the rows, dwarf beans may be planted between the celery rows already mentioned, but *only two beans to the hole.* In very good soil onions and early lettuce can be put together. The lettuce is harvested and the onions spread out. Although onions grow poorly in sand, they manage to get along in it quite well if camomile has been sown thinly in between.

When sowing legumes—peas in the garden, or lucerne in the field—it is preferable that they be planted in double rows. These plants will give one another mutual aid in their growth and will still have on the outside, between them and the next double row, enough room for development.

It is not our purpose here to present a gardening text-book. Anyone wanting to make use of these suggestions ought to be an experienced gardener to start with. Rather are these references intended to direct attention to problems which have not heretofore been given enough observation and study.

The question of manuring is of particular importance where intensive use is made of garden land. Until now we have treated those aspects of this question which deal with soil conservation and which avoid methods resulting in impoverishment of the soil.

The care and treatment of manure has already been discussed

SOIL FERTILITY, RENEWAL AND PRESERVATION

in Chapter VI. But for garden purposes there should be an intensification of this treatment. We shall only outline the principle here. What is done with it must be left to the individual. As a first instance, raw, strong, intensive manuring, especially the use of liquid manure, forces the plants to a heavy growth. They produce thick green leaves. When such vegetables are cooked, the odour coming from the kitchen tells us just what sort of manure has been used in their culture. Such variations of odour arising from the differences in the manures used are very obvious as already stated in the case of cauliflower, for example.

In such raw, immature manure, the strongly smelling products of partially disintegrating albumen and nitrogen compounds have not been completely consumed, but have been taken up by the plants directly from the soil. Apart from their odour, these compounds also have certain harmful incidental effects on the living things in the soil, and on the plants themselves (promoting susceptibility to fungus diseases) and, when eaten, on the human organism. The result is digestive disturbances, producing heartburn, bloating, and if the process goes far enough even causing infection with intestinal worms. These worms are to be observed in the muck when the night-soil fields near the cities are drained. Many stomach and intestinal disturbances are cured without treatment when the use of vegetables fertilized in this way is discontinued.

Two aspects of the question to be emphasized here are: (a) intensive fertilizing through the greatest possible conservation of manure values (cf. Chapter VI); (b) the needs of hygiene and health which demand the best taste possible and the very best *quality* in the products of the garden. The more finely rotted, the more completely transformed into odourless humus a manure is, the better is it suited for gardening. The more this manure is mixed with compost, the finer and more aromatic are the vegetables.

Thus we see the need of manuring with prepared, well-rotted stable manure as the basis for early potatoes and other plants with heavy soil requirements; and when we wish to grow an especially delicate crop special manure should be prepared for it. The simplest method is the mixing of half-rotted manure with half-rotted compost. Both heaps may be located near one another and, in turning them, the two heaps may be combined in alternate layers (at which time the preparations may then be again inserted), or fresh manure may be added to three-months-old compost. There

COMMENTS ON MARKET GARDENING

are many possible combinations. This composted manure (manure alone may also be mixed with alternate layers of earth) or mixed compost is suitable for all finer plant cultures, for greenhouses, etc. When it has completely turned to soil, it may be put in small amounts directly into the plant hole or the furrow prepared for any kind of plant. This procedure makes possible its complete utilization by the plant. This high quality manure-compost is as suitable for potted flowering plants as for fine vegetables.

In practice we first prepare the seed bed, then use bio-dynamic preparation 500 directly on the soil, in the open furrow or the planting hole, putting in some compost, and then sow or plant. When transplanting it is good to give the roots a bath in a dilute solution of preparation 500.

For cucumbers a compost with a heavy proportion of manure is advisable. These, like tomatoes, may be planted on high ridges or on little mounds of earth. A handful of compost earth is then put in every seed or plant hole. It may be added that cucumbers and tomatoes do poorly together. A suggestion of Dr. Steiner in regard to tomato culture has proved itself particularly valuable in practice. While in general it is better not to grow a plant in a compost made from its own remains—this is especially important for cucumbers and also for cauliflowers and other cabbage varieties—Dr. Steiner called especial attention to an exception to this rule in the case of the tomato which is particularly at home when grown in its own compost. All the tomato refuse, leaves and stems, is set up in layers with earth according to the familiar composting technique. The compost is then ripe in time for the next season and is put into the planting hole. For greenhouse culture of tomatoes it is especially important to have good ventilation and dry air. Moist air under glass promotes the development of the fungus pests. Underground irrigation has been found valuable here. Pouring water on the plants always brings dampness into the air; underground watering lets the plants have the necessary moisture and yet keeps the air dry. Perforated pipes are laid at a depth of about ten inches at a slight incline and parallel to the plant rows. These are provided with water from a central channel. The amount is regulated so that the moisture is just visible at the surface of the ground. This procedure will not work in a porous, purely sandy soil, because of the great water loss such ground entails. With this exception, the method has proved itself valuable everywhere, and above all has been helpful in the fight against fungus pests. With such irrigation and the use of tomato compost and the other bio-

SOIL FERTILITY, RENEWAL AND PRESERVATION

dynamic procedures, we have been able, for seven years now, to have full yields from an annually repeated tomato culture in a 1000-square-yard greenhouse, without any plant diseases.

In the first year there was still some blight and leaf curl, later there was none. It was found necessary to grow our own seed. We observed in our gardens that only with our own seed were we in a position to keep the plants permanently healthy.

The tomatoes we harvested were 87 per cent first quality (the most saleable, commercial export size), 7 per cent of B quality, which were larger, and 6 per cent of C quality, smaller tomatoes. This was the figure year after year. Much attention was given to keeping the plants evenly trimmed to single stems. An even, healthy growth was in fact one of the most important results of this method of using fine quality compost.

These special composts are capable of every possible variation. To describe them all here would take up too much space. But it should be emphasized once more that cauliflower compost should *not* be used on cauliflower plants, *nor* cucumber compost on cucumbers. Composted pig manure is especially good for leeks and celery. Composted chicken and pigeon manure can both be put to good use, for example, in horticulture, and in all situations where a strong forcing effect is needed.

We come now to a chapter of gardening which, we regret to say, has become very important—the subject of pests. Instead of using poisonous copper, lead and arsenic preparations, the pests can be attacked in a biological way. For this purpose the life rhythm of the plant and the pest which has attacked it must be studied. The presence of plant lice on broad beans is a very instructive example of this. They attack the beans at a definite point of their growth, that is to say, from the forming of the fourth leaf group to the development of the eighth. If weather conditions are helpful, the plants eventually get over the attack. If the contrary is true, they turn black and die off. Careful observation here can teach us a great deal. The attack of plant lice becomes especially strong when there is not sufficient air circulation about the beans; this is the case when they are planted too thickly. It is also strong when cold or drought suddenly checks the growth of the plants. The beans should be planted in series of a few rows, and then their growth should be observed very closely. When the danger point is reached they should be sprayed with preparation 501, which stimulates plant assimilation and upward growth. This helps the plant to grow swiftly past the critical point and to reach sooner the later

PLATE 4. Normal, healthy tomato plant from a greenhouse treated bio-dynamically for seven years. Height of the whole plant, six and a half feet

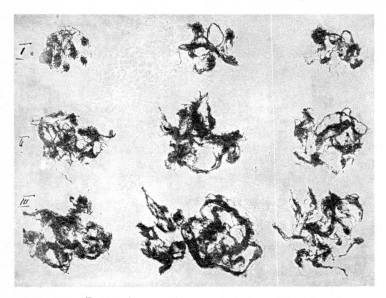

PLATE 5. Roots in another experiment like the first. I. Was with copper and lime mixture. II. Without any special treatment. III. With the clay, manure and sand mixture

COMMENTS ON MARKET GARDENING

stage at which the sap develops a different taste no longer agreeable to the plant lice. (Reviser's note: Tests in the United States, utilizing the same procedure to help dwarf beans against the Mexican bean beetle, have shown most satisfactory results.)

Another example is furnished by aphides on fruit trees. **Dr. Steiner advised planting nasturtium among the trees.** We have carried this suggestion further, not only making the plantings but carefully painting and spraying the trees with an extract of nasturtium. Such plantings are effective because the nasturtium contains a strong aromatic substance which also penetrates the ground through its roots. This the trees can take up through their roots and bring into their sap stream, making changes in it which are imperceptible to us but very evident to the fine organs of smell and taste possessed by the insects. Result: the aphides disappear. Besides this, of course, we should not forget that the proper general treatment of fruit trees is of great importance in helping to restore their powers of resistance to pests.

Another observation may be made concerning flea beetles. These are driven away by shade. They prefer a crusty soil, the surface of which, swept by the wind, becomes dry and impermeable. Hence we should work to develop a crumbly soil with strong capillary action, and to make a shade by means of mixed crops and catch crops. We should also mulch the soil between the rows with partially rotted leaf compost. The flea beetle shuns tomatoes and wormwood. It is therefore advisable to plant these at random, here and there, between cabbage or radish plants. Even the spreading of the trimmed-off shoots of tomatoes can be helpful.

Another pest we have to fight is clubroot. Its development is promoted by the use of unrotted, raw manure; or of unmixed manure, such as pig dung only, or goat dung only, or of raw, uncultivated, or poorly worked soil, or of too little manure. Large amounts of ripe, mild compost and later also compost from herbs and vegetable plant refuse put into the planting hole can be helpful in overcoming this condition. To make possible the early planting out of cabbage seedlings, we can use some leaf or straw compost as a moisture-holding mulch round the plants.

Since the cabbage worm and butterfly are repelled by hemp, tomatoes, rosemary, sage or peppermint, it would be wise to use such plants as protective catch crops. The asparagus beetle is repelled by tomatoes, the mole cricket by hemp, birds by a pickled herring hung from a pole. Birds also are very "skittish" and shy when confronted with the decaying body of a bird of the same species hanging from a pole. Scarecrows fulfil their purpose only when they are moved about frequently. Strips of any glittering material hung so that they are in continual movement make a most excellent "scare-crow".

To combat the capsid bug the blossoming tree must be given a highly diluted liquid spray of slaked lime. It is often also advisable to sow or scatter the seed of "attracting plants" to entice insects from plants needing protection. In this way we can use lettuce, spinach and potatoes to combat the cockchafer larvae, strawberry weevil, woodlouse, bollworm. There are many possibilities for helping oneself in such natural ways. For instance, in order to catch snails the empty half orange or grapefruit skins may be laid between the bed with the open side down; after a certain length of time a whole collection of these pests will be found under them.

For the care of fruit trees there are some special rules. Biologically, the fruit tree stands between the woodland tree and the cultivated field plant. It needs care, but no intensive fertilizing. It has in any case a longer cycle of growth than the annual or biennial. The four chief causes for trouble in an orchard are: (a) too strong fertilizing, especially if this has been done with fresh, raw manure; (b) too thick a stand of trees, which permits too little light and too little movement of air; (c) the wrong tree stock, for the tree in question; (d) and lastly, the use of a variety of fruit tree on a soil and in a climate to which it is not suited. These four points must be considered by those who aim to improve the general health of their orchards. Obviously results cannot be obtained by working *contrary to* nature. If the wrong rules of procedure have been followed and the constitution of the tree has been weakened, the impossible should not be expected.

One thing should be remembered above all; only the best, rotted manure and compost helps the tree. Only when this is used does the fruit ripen well, stay on the tree instead of falling too soon, and retain good keeping quality. Too intensive fertilizing or use of raw manure is sure to bring the opposite to pass. Here the relationship with the forest tree may be seen. Around the tree a firm, matted, mossy, crusty layer of soil should never be allowed to form. The roots of trees, like those of other plants, prefer a loose, airy soil. If the tree stands in a meadow the soil must be loosened from time to time by digging up under the overhang of the widest limbs and branches, and then there should be a light fertilizing of the soil with prepared compost and preparation 500. It is useless to put fertilizer close to the trunk because here there are no fibrous roots to take up nourishment. Where the field is to be used exclusively as an orchard, it has been found valuable to keep the ground open for two years, digging it up as often as necessary, then covering it afterwards for two years with a legume, such as clover. This latter can be used also as fodder or composted. In this way provision is made for natural nitrogen fertilizing and for a loosening of the soil. In this way the fruit tree is not sated with N_2, a substance to which it is extremely sensitive. Among our special measures against fungus diseases we include spraying with preparation 508 (Equisetum arvense). We use this in fruit growing by painting it on the trunk and lower limbs of the trees and by spraying the crown. This must be repeated as often as necessary, and is most effective when begun early as a prophylactic treatment.

Special benefit has also been found from painting the trunk and spraying the crown with a mixture of one-third clay, one-third cow manure, one-third sand and as much water as is needed to make the mixture thin enough to paint or spray. In the water either 500 or 508 may be used. This is applied in the autumn and repeated if necessary before sprouting time in the spring. It stays for many months on the trees and helps the formation of a healthy, dense, enclosing bark. It stimulates the cambium layer, heals wounds and stops bleeding of sap, and in general it has shown itself to be of great benefit to the health of trees. Cankers, cleanly cut out and painted with this mixture, heal with smooth edges. The use of tanglefoot (rosin and castor oil) should never be forgotten in the case of trees. In the early spring the bark of the tree should be thoroughly cleaned for several inches about three feet up from the soil. The cleaned part of the stem should then be covered with the liquid glue material and upon this should be laid a clean strip of cotton cloth (linen) also covered on the *outside* with the tanglefoot glue. This is the best trap for all beetles migrating from their winter rest in the soil to a home in the tree.

The following experiment has been tried with the clay, cow manure, and sand paint previously referred to: groups of flower pots were painted with (a) a copper and lime mixture; (b) a carbolic acid solution; (c) the above clay, manure, and sand paste; (d) nothing, as a control. Tradescantia shoots of equal size and age were set in a mixture of sand and humus in the pots. After a period of time the plants were dug out and the development of the root

SOIL FERTILITY, RENEWAL AND PRESERVATION

weight was determined. The paste (c) had had an especially strong effect on the development of the root weight.[1]

It is advisable to wash down the trees with preparation 508 before using the mixture, having first brushed any fungi and algae off the trunk. An objection is heard—"That makes more work!"—as if the great amount of spraying with every sort of material in recent years did not also make work. The work of the two procedures balances fairly well, it is about the same in both cases, but with this method we have the satisfaction of knowing that we are working in a healthy, natural fashion. The same treatment has also proved its value for grape culture. Grapes need in addition an intensive spraying with preparation 508 during the critical time of the early summer. With fruit trees and also with berry bushes, a light compost fertilizing in the late summer is advisable, at the time when the fruit is already well formed and the leaf masses are still green. The set of the fruit for the next year is helped by this treatment. For strawberries this is done in August after the harvest in order not to harm the fruit. This fertilizing must not be strong enough to have a forcing effect. Although, in order to complete a picture of the bio-dynamic method, a whole series of detailed procedures still remains to be outlined, it will not be possible to complete this in the space of so small a book. It is better that we leave these questions to be answered as they arise by means of advice from bio-dynamic information centres.

[1] Average figures for 16 plants in each case.

	Weight of the shoots at planting	Weight of roots in grammes at conclusion	Weight of green plant parts
a.	2.87	3.51	35.25
b.	2.58	3.11	43.17
c.	3.0	6.62	46.8
d.	3.18	4.11	49.12
A second experiment:			
a.	5.02	6.35	17.23
c.	4.72	8.73	30.8
d.	5.05	4.97	28.62

CHAPTER ELEVEN

The Dynamic Activity of Plant Life—
Some Unaccounted Characteristics

In studying the construction of a living thing, we first observe its purely material side, its weight, its shape, its chemical constituents, its structure. These material elements are combined, are quickened; it grows, reproduces itself, takes definite forms, temporary or constant, undergoes change of substance: in short, it shows biological formation, it has a living development and order.

But knowledge of its chemical formula is of itself insufficient to enable us to understand its physiological reactions. We think ourselves nearer comprehension when we speak of its wholeness, its entelechy, when we regard it as a higher unity of organic differences.

The question, "Why does a given reaction take place just at this place and time?", leads us beyond a consideration of the purely static condition to the recognition of organized, directed processes —i.e. dynamic happenings. Dynamic, from the Greek word "dynamis" meaning power, is applicable to everything in the interior of a life process which gives this a definite direction, activity and order: that is to say, everything which causes this life process to be influenced, assisted or checked. Over and above the material, static existence we see the life condition with its possibilities of development or decay. That which promotes or impedes this condition, above all that which first renders it possible, we call "dynamic activity".

Light and warmth must be named as the first promoters and awakeners of life. Without heat, there can be no life. Light assists growth, furthers assimilation and—as we shall see later—has an appreciable influence on form. Chemical forces work in and through physiological reactions. In the living organism these are directed to an end, they do not proceed at random. The organism makes a selection between possible reactions and thereby proves the existence of wise laws to which it conforms.

Already in the inorganic we see directing, helping or retarding processes. In chemistry these are known by the generic term "catalysis". A catalytic substance aids a chemical change but remains itself unchanged in the process: that is to say its *presence* is sufficient for the effect.[1]

In life development we find a number of such catalytic, i.e. directed processes: we therefore speak of "bio-catalytic" substances. It is not only inorganic compounds, such as metal, hydrogen, etc., which can release dynamic activity, but also organic, often very complicated, substances. Their presence is shown in the physiological course of events, in the quickening or slowing down of growth, in the assimilation of nutriment and in respiration. It can also be discerned in alteration of form, in one-sided development of height or thickness, of foot formation, of ripening, etc. Substances producing such effects are called by different names, according to the "dynamic tendency" they display: growth-promoting, i.e. auximone (auxin), hormones, growth hormones, ferments, enzymes, vitamins. "The determining factors of form, however, include more than catalysis. Catalytic substances ensure that something happens but have no share in deciding *when* it happens." (Mittasch.)

In what follows we aim at giving a summary review of the most important "dynamic activities" in the plant world. In our view, the most striking quality common to all those substances which exercise influence upon their surroundings is the fact that all such bodies as come into question are present in a highly rarefied form and that this rarity is frequently a necessary condition for the effect produced. This indicates that the chemical reaction is unimportant in itself and soon vanishes, that through it, however, a whole set of phenomena is initiated. The change from tadpole into frog is influenced by applying extract of thyroid gland; a rapid transformation into very tiny frogs results, the hormone producing this result when in a dilution of one to one thousand millions.

Fundamental in the study of this subject was the observed fact that the "air exhaled by apples" affected the growth of neighbouring plants. In this exhalation ethylene is found. True, the quantity

---

[1] "The catalytic substance as the simplest directing factor is, and in the modern dynamic—no longer mechanistic—view of the world remains, a Cause, not productive of energy it is true, but nevertheless an important cause. . . ." (Alwin Mittasch.)

THE DYNAMIC ACTIVITY OF PLANT LIFE

is small; an apple generates in some months of ripening only 1.26 mg. of ethylene. Nevertheless the effect is powerful. Molisch[1] noticed amongst other things that plants in its vicinity reached maturity earlier and lost their leaves sooner, and that the exhalation was injurious to the growth of blossom and of root.

Boas[2] writes: "If, for example, broad beans, sunflowers or pumpkins are left to germinate in earth or in damp sawdust under glass near apples, the exhalation of ethylene causes strange transformations. The pumpkin shoot resembles a little kohlrabi. The sunflower bud is like a diminutive sow's tail. The broad bean sprouts (Vicia Faba) let their normally upright shoots creep along the ground. Thus the normal response to gravity, the upright position in space, is destroyed."

The germination of rye is hindered by apple aroma in the dark, but furthered in the light. (This fact leads the writer to wonder whether, if apples were laid near potatoes in a cellar, the latter would be kept from sprouting, if for example the apples and potatoes were laid one above the other and the smell of the apples was shut in. This of course presupposes apples of a good keeping variety of a sort to last until spring, a sort which is increasingly rare.) Dandelions also exhale ethylene and thereby further ripening.

These dynamic activities are put to technical use in the artificial ripening of fruits plucked unripe, such as oranges, lemons, grapefruit and bananas, which are "gassed" to make them appear ripe.

Following upon his countless observations of the most delicate mutual effects amongst plants, Boas insists on the need for a dynamic Botany, which in addition to the static descriptions (Morphology, Physiology, especially Physiological Chemistry, Anatomy, etc.) shall treat of these effects.

"If we deal with the plant world from the point of view of effects," he writes, "there emerges a new behaviour of plants. Influences are everywhere at work, all affect each other. Life in its circulation produces an invisible entanglement of effects. To throw light upon this is the task of dynamic Botany."[3] "There is no doubt that our present-day Botany fails to grasp the entire plant. . . . Inadequate dynamic knowledge of the plant world means in every case biological uncertainty. Because this state of uncertainty

---

[1] A famous Vienna plant physiologist.

[2] Dr. Friedrich Boas, *Dynamische Botanik*, Verlag J. F. Lehman, Munich, 1937.

[3] F. Boas, op. cit., pp. 11–12.

has in fact existed for centuries it does not follow that we must continue to submit to it."[1]

The growth of yeast can be influenced by dynamic substances. There are yeasts which do not sprout even in favourable media. Only when certain plant juices (or humus) are added, even if only a trace, do they begin to grow.

The presence or absence of a bit of filter paper may, under certain circumstances, influence a process. The addition of filter paper, for instance, will promote the rate of cell transformation. This is related to the fact that through the presence of the filter paper a higher degree of acidity is reached and speedy changes follow. Insight into the chemical details of this process is still lacking.

The agricultural chemist is wont to refer to potassium as an important foodstuff, and to demand that it shall be present in such and such percentages or, if not, that its absence be made good. Unfortunately, he tells us less about the dynamic activities of this potassium. For example, reduction of (not yet lack of) potassium tends to heighten assimilation and transpiration, i.e. tends to an increase of leaf green (chlorophyll) content. An excess of potassium makes the assimilation decrease. The ratio between the nitrogen in the form of albumen and the nitrogen in the form of nitrates is controlled in the plant by the potash content. The same applies to the formation of starch.

An experiment in manuring a certain kind of pear tree showed that the content of malic acid in the fruits was greatest when well manured, least when potassium was absent or when the ground was left unmanured.

Boas writes[2]: "It is clear from the figures given that hundredweight does not equal hundredweight, that albumen is not the same as albumen. The physiological-dynamic values cannot be supplied from statements of the water, sugar, albumen, fat and ash content. They require special research with new methods . . . then, together with the old tried methods of analytical chemistry (sugar, albumen, fat, ash), we may arrive at a real delineation of plants and their ever-changing life processes."

A dynamically interesting plant is the crowfoot or buttercup. In its green condition it contains a substance which checks growth. This crowfoot poison (anemonin), extracted in small quantities from a fluid secreted by the plant, prevents putrefaction and kills fungus and bacteria. Crowfoot juices remain for weeks unfer-

[1] Boas, p. 10.   [2] Boas, p. 40.

mented; added to other juices they check fermentation. Their effect in a dilution of 1:60,000 and 1:250,000 can be clearly observed; in fact, they were effective in a yeast culture at a dilution of 1:33 thousand millions, and that was not necessarily the limit of their effectiveness. In a dried condition of the plant, however, this growth-checking substance is not found. Instead there is observed a growth-promoting substance.

What does this signify in practice? Take a meadow containing a quantity of crowfoot. Its active substance will through rain and decay percolate into the soil; there it checks the life in the soil, "sterilizes" it, checks the formation of humus. The meadow becomes sour, other plants die, especially clover, and next year there is an even more favourable soil for the crowfoot. Curiously enough, kaolin and clayey soil absorb this harmful stuff. Therefore thorough manuring with clay compost, together with a good aerating (harrowing), is an excellent counter measure. It is important to note how the crowfoot action appears even when the substance is in highly diluted form.

Camomile has the effect of expediting growth in yeast cultures and also in the life of the soil. An extract of 0.0012 mg. of camomile applied to a yeast culture increased its growth about four and a half times above that of a similarly grown culture without the camomile. An extract of 0.003 g. increased it 110 times. The limit of the active power of camomile was worked out by Boas at 1:125 milliards. A decade ago such activities would have been decried as mystical; to-day there are thousands of scientific writings embodying the results of countless observations of them.

In stable manure the same kinds of powerfully active substances may be found, whose presence or absence may be decisive for physiological development. So may the presence or absence of retarding substances. Amongst other things these affect root formation. "Hence it arises that the growth of roots may be actively promoted by a bath of plant-extracts. This proved valid in the case of bean-roots which were steeped for a short time in a decoction of camomile (10 g. in 500 cc. water). Individual plants and parts of plants produce different blends of active substances."[1]

The writer has demonstrated the value of bio-dynamic preparations added to manure and their effect upon the growth of roots by countless experiments. In these preparations it is partly a matter of plants like camomile, dandelion, yarrow, etc., which have been subjected to slow fermentation. The growth-promoting substances

[1] Boas, p. 73.

thus set free in them can be applied to the soil in a compost or a manure or directly by dilution. They increase fermentation and augment the life of the soil. Knowledge of their chemical formula is of minor importance; it tells us nothing concerning their activities. For more than a decade these preparations have met with criticism. But I believe that no one who follows the new studies concerning the so-called active substances can keep his mind closed to the effects of the bio-dynamic preparations. An objection frequently made to these preparations was founded on the high degree of dilution in which they were applied. As compared with the very high dilution of the growth-stimulating substances, auxins, etc., the bio-dynamic preparations appear rather "concentrated" when applied, e.g. in a 0.005 per cent solution. According to Professor Kögl of Utrecht, the well-known auxin expert, the yeast extract biotin works in a dilution of 1:400,000,000,000.

To return to the effect of camomile, the writer proffers an observation of his own. The active parts of the plant naturally make themselves felt wherever the plants grow and through root secretions or the falling of leaves influence the soil and the roots of other plants. In the present book, he therefore writes of "dynamic" plants. Only those who ignore the new plant physiology can disapprove of this conception. It has been observed that camomile deliberately sown as "weed" amongst wheat or other cereals influences their growth. For instance, a large quantity—20 camomile to 100 wheat plants—when sown together results in the production of small and light seed. But a smaller quantity, for instance 1 camomile to 100 wheat plants, results in a large and heavy grain. Only in the greater "dynamic" dilution can the substances do their work.

A similar substance, related to yeast extract, is found in the sap of the birch tree. It especially affects fungus. This substance may also be secreted in the roots of the birch or be washed by rain from its trunk into the soil. This explains, first, Dr. Steiner's statement (1921) that the soil surrounding birches is found to invigorate growth, and, secondly, the fact that composts under birches rot into finer earth with less smell than elsewhere. For these reasons, we recommend the planting of birches on or near compost and manure yards. These and elder trees, which have a similar effect, will send their roots into the compost soil, exercising their beneficent influence. Experiment will show that such composts rot more quickly. Of the same order are the important effects of couch grass, blackberries, elders, nettles, which play so great a part in the afforestation of unfavourable regions. If some day the effects of the root secretions of these plants are more closely studied, much material for research will be found therein.

We are here moving in the sphere of purely dynamic events, which were strange to us fourteen years ago but now are familiar through numerous scientists and hundreds of experiments. Whether by attributing "dynamic" processes to special substances we have solved the problem of their activity is another matter. To refer an activity back to a chemical, perhaps precisely defined, substance is no real explanation, for the formula throws no light upon the activity. We only know that the substance functions as its bearer. Through it, effects are wrought which, for example, resemble an increased, diminished or one-sided effect of light, warmth, gravity or chemical affinity. Further study will be devoted to this problem, which is but a section of a wide new field of research.

We conclude our introduction to the idea of dynamic forces with these words of Boas: "At all events, the great dynamic importance of the substances which impart growth-influences will persist, whatever individual knowledge may yet be gained concerning them. And this importance will remain valid for the course of their activities. For they are chemically constant—and remain effective in nature when they are detached from the plant and, for instance, penetrate the soil. Perhaps they also carry elements of fruitfulness."

The nourishment of a plant consists in the assimilation of salts, of water, and of carbonic acid taken from the air. Its total mass is composed to the extent of 90 per cent and more of water. Only from 2 to 5 per cent of its mass comes from salts taken from the soil. Thus it is a surprising fact that *the plant receives an important part of its nourishment from the air* which we can neither fertilize nor influence in any way—*and only a relatively small amount is received from the soil* which we *can* and *do* influence and fertilize.

Experiments, for example, which consisted of increasing the CO_2 content of the air in greenhouses, have had no practical significance. We have little influence on the volume of nutritive material coming to the plant from the air. Only at one point can the absorption of carbonic acid be strengthened: by increasing the so-called soil carbonic acid. Considerable research has shown this soil carbonic acid to be an especially active and beneficial agent in plant growth. Soil carbonic acid is produced by the micro-organ-

SOIL FERTILITY, RENEWAL AND PRESERVATION

isms in the soil, and thus depends upon the degree of humus and microscopic life in the soil itself. Here we can influence the plant's nourishment "coming from the air", in connection with one of the most essential nutritive materials.

These carbonic acid salts may be taken up directly by the roots if the salts are present in a soluble, that is, in an absorbable form. It is, therefore, to the farmer's advantage to bring his soil into the proper condition. This is accomplished more easily when there is considerable water in the ground. We must, therefore, look after the soil's water supply in dry regions. Humus and a crumbly structure are important factors in this connection, as we have frequently pointed out. They help in the process of weathering, and of making available the mineral substances in the ground. This process is also advanced by the plant roots themselves, in part mechanically and in part by the effect of their own secretions.[1] These properties of the plant must definitely be taken advantage of in a rational agriculture. Any sort of treatment that enlivens the soil, that makes it richer in humus and opens it up and loosens it, we define as "biologically effective". Ways to attain these ends are described in this book.

Much is said about the chief nutritive materials, potassium, nitrogen, phosphoric acid and calcium. According to the law of nutritive substances, the volume of these materials taken out of the soil by a harvest must be replaced by fertilizing. This law, in itself correct, logical and proved by laboratory research, is subject to constant variations under natural, outdoor conditions. In practice, we do not know exactly which substances are taken away, and in what form; nor do we know which, on the other hand, are replaced by weathering in the soil, being carried back in atmospheric dust and in water. Nor do we know whether a specific method of working the soil cuts off the access of such materials or brings them into the constant cycle of substances used and replaced. Even when we fertilize with a substance we may disturb the balance and cause deficiencies in other salts.

We have called attention to the rhythmic variations in the solubility of the phosphoric acid compounds. Put briefly, we have in them a series of biological phenomena the course of which is very hard to control, but which is as important for the nourishing of the plant as the purely quantitative "replacement". States of balance or of unbalance which must be taken into account are con-

---

[1] Cf. discussion of the effect of various plants in making soil materials available, ch. vii, p. 68.

THE DYNAMIC ACTIVITY OF PLANT LIFE

stantly arising. There is, for example, the nitrogen problem. It is plainly evident that there is an antagonism between the nitrogen produced by bacteria and the nitrogen coming from the use of mineral fertilizers. Legumes "forced" with artificially applied nitrogen develop no bacteria. Clover on meadows fertilized with ammonium sulphate disappears. Fresh liquid manure (with its free ammonia) has the same effect on clover as the ammonium sulphate.

An experiment carried out in Holland[1] gave the following results:

On an experimental field of six groups of ten parcels each, ten received no nitrogen before the first cutting, ten were given Chilean nitrate, ten calcium nitrate, ten ammonium nitrate, ten calcium ammonium nitrate, ten ammonium sulphate. The application consisted of 40 lbs. of nitrogen per acre. Every year the mowing was done early, that is, at the best possible time for mowing, and for the rest of the season the fields were pastured. The clover growth on the untreated parcel proved to be the strongest.[2]

The hay yield on the unfertilized parcel was quantitatively smaller (because of its high protein content), yet in spite of this small yield of hay the total protein yield was greater. Frequent nitrogen fertilizing drives clover out of a field; late mowing and irregular grazing do the same.

The following figures in lbs. per acre of experiments on the development of weeds in meadows fertilized with nitrogen fertilizer are also interesting:

Nitrogen fertilizer in increasing doses	0	60	90	120	150
Amount of Crowfoot (Ranunculus acris) in clover fields	0.5	1.5	3	4	5
Clover harvested	7	4	4	3	2

This means that clover parcels without mineral nitrogen harvested 3.5 times more clover than the clover parcels fertilized with 150 lbs. of pure mineral nitrogen.[3]

---

[1] De Boerdery, 24 July 1935.

[2] Analyses of the hay showed the following:

			Percentage of protein	Percentage of lime
Parcel without nitrogen fertilizing			12.9	1.33
,,	with	Chilean nitrate	8.9	0.75
,,	,,	Calcium nitrate	8.5	
,,	,,	Ammonium nitrate	8.7	
,,	,,	Calcium ammonium nitrate	8.7	
,,	,,	Sulphate of ammonia	9.2	

[3] Cf. ch. xiii, p. 163. Strong mineral applications of nitrogen lower the protein content, according to Prof. Boas of Munich.

We see plainly from this example that the effect of nitrogen fertilization is not only quantitative but that it affects the whole biological process.

In this connection, we can include another important observation. The mineral substances in the soil are in a state of equilibrium. When, however, a soluble material is strongly preponderant, then other materials are driven out of the solution, precipitated, etc. It is well known that a heavy potassium fertilizing precipitates the soil magnesium. Researches have shown that on most agricultural soil there are continual disease conditions of rye and oats, and it has been proved by analysis that these have been caused by a lack of magnesium. Prof. Schnitt's research leads to the view that, where there is strong acidity, the magnesium is so vigorously absorbed by the soil that it is no longer in a condition available to the plant. The precipitation of phosphoric acid in sour soils and its fixation by heavy soluble iron and aluminium phosphate seems to be hindered by the presence of magnesium. It is suggested that the problem of maintaining the magnesium content of the soil is difficult only because there is such a variety of factors working together here. For instance the activity of magnesium is of decisive importance for chlorophyll building, for the development of protein and the utilization of phosphoric acid in the plant. The various factors mentioned above include the condition of the soil, its magnesium content, its acid or alkaline reaction, its content of calcium and of other salts, humus and the effect of the plants growing on it. "The absolute magnesium content of the soil is of minor importance when the question arises as to whether there is need for magnesium fertilizing. Much more important is the knowledge of the degrees of magnesium tolerance, which may be so low in the case of sour soils that they eliminate any added magnesium by precipitation to such a degree that the added magnesium produces no effect. Because of the strong effect of magnesium, small amounts are sufficient to produce large results," says Scharrer.[1] We learn that the important thing is not at all the absolute volume, but the condition of the soil. Conditions can arise which figuratively may be likened to a thirsty horse, tied to a post, near a spring, with a too short halter.

Recent developments in plant physiology have taught us that,

---

[1] Prof. K. Scharrer, Tech. Hochschule, Munich, "Bedeutung seltener Elemente für die Landwirtschaft", *Mitteilungen für die Landwirtschaft*, 1937, Vol. II.

THE DYNAMIC ACTIVITY OF PLANT LIFE

in addition to the coarser substances present in plants, there is a whole series of ingredients, often present only in extremely minute quantities, which are absolutely essential to plant life. These substances were not considered in the fertilizing and yet they are present. The role of boron is familiar. Legumes only develop root nodules when there are traces of boron present. Solanaceae like tomato and tobacco, and the chenopodiaceae grow healthily only where there are traces of boron. In the case of the sugar beet, the rotting of the core can be cured by spreading over the field from 4 to 13 lbs. of borax per acre.

The question of copper is also illuminating. Copper sulphate in a dilution of 1:1,000,000,000 injures algae (spirogyra); in a dilution of 1:700,000,000 it hinders the development of wheat sprouts; at 1:800,000 it stops their growth. In the soil it is still a strong poison to bacteria in a dilution of 1:100,000. Nevertheless various crops contain large amounts of it: oats up to 0.9 per cent, barley, wheat and rye up to 0.01 per cent, potatoes 4 milligrammes per kilogramme, hay from 6 to 12 milligrammes per kilogramme, lentils 0.015 per cent, broad beans (Vicia Faba) 0.03 per cent, peas 0.01 per cent, soya beans 0.01 per cent of the ash weight. It is found in oranges, blossoms of the European elder (Sambucus nigra), water melon, gourd seeds, black mustard, maize (corn), pine (wood and bark), German iris and other plants. This copper is found in the plant without the soil's ever having been treated with it, even when it is scarcely possible to find a trace through soil analysis. Here we have the capacity of the plant to gather the most minute quantities of certain elements and "extract" them from their environment.

There are also plants that accumulate lead, such as: Festuca duriuscula (containing up to 12.25 per cent of Pb_2O_5, found in ash) and Randia dumetorum. Titanium is found in the wood of the apple and pear tree, and manganese is of significance for the grape vine. Iron is stored up in many plants up to high percentages in the case of the Acacia cebil, or the pine. Hugo de Vries in his classical *Leerboek der Plantenphysiologie*,[1] presents some analyses of aquatic plants which make these characteristics very clear. He says that "the chemical combination definitely does not conform to that of the soil or the water in which the plant grows. And sometimes variations found in two plants growing close together are very great." Thus, for example, the analysis of

[1] Haarlem, 1906; p. 201.

SOIL FERTILITY, RENEWAL AND PRESERVATION

certain plants growing in the same ditch shows the most varied results.[1]

The capacity for absorbing such substances in highest dilution and then accumulating them belongs in common to all the kingdoms of nature. A whole series of substances, especially the heavy metals, are present in the human body also. If these are present in our food, only the slightest traces of them can be discovered or none at all. The new, refined techniques of chemical analysis have only lately given us the possibility of investigating these things. I am reminded in this connection for example of the work of Dr. J. Noddack of Berlin.[2] Refined spectrum analysis shows that one can speak of the *universal presence* of any substance, and in general of every substance, including thus also the heavy metals in a dilution of from 10^{-6} to 10^{-9}. Certain organs of plants, animals and man can accumulate these substances in measurable amounts. It only remains for research to discover how this takes place.

Here is a task the solving of which would lead to an appreciable enrichment of our knowledge of the phenomena of life. Hardly anything in nature is present without reason and purpose. Thus it may, perhaps, be possible even to-day to say hypothetically that these minute quantities of substance in the plant world have a guiding and directing task to perform in regard to the various biological processes that occur. It is already possible from the study of the substances existing in large volume in the plants to see that their presence gives direction to plant growth. The author is thinking here, for example, of the influence on starch formation resulting from the absorption of potassium by the plant, on protein formation resulting from the absorption of phosphoric acid, and

[1] The percentage of various minerals found in plant ashes:

	Chara foetida I	Chara foetida II	Hottonia palustris	Stratiotes aloides	The surrounding water contains the substances in the following percentages
Potassium	0.49	0.23	8.34	30.82	0.00054
Sodium	0.18	0.12	3.18	1.21	—
Calcium	54.73	54.84	21.29	10.73	0.00533
Magnesium	0.57	0.79	3.94	14.35	0.00112
Phosphoric acid	0.31	0.16	2.88	2.87	0.00006
Carbonic acid	42.60	42.86	21.29	30.37	0.00506
Silicic acid	0.70	0.33	18.64	1.81	traces

[2] J. and W. Noddack, "Herkunftsuntersuchungen", *Angewandte Chemie*, 1934, No. 37.

on chlorophyll formation resulting from the "presence" of iron, in spite of the fact that in the latter case iron is not contained in the chlorophyll molecule, but exists only in the environment. It has been shown that not only a lack, but also an overabundance of substances may be harmful here.

The presence of minute quantities of this or that substance has apparently more a functional than a nutritive significance. The total functioning of the organism is in many respects dependent on the status of these finely atomized substances. The health of the plant is influenced by them.

However, modern research goes a step further. It now recognizes "chemical effects at a distance". Dr. Ried in Vienna, Prof. Stoklasa in Prague and Councillor v. Brehmer in Berlin, by means of various methods of research, have obtained the same result, showing that the presence of mineral substances, even when removed to a certain distance, can have considerable influence on the growth of the plants.

Brehmer[1] reports that potassium in the vicinity of potatoes—separated by an air space from the containers in which the plants are growing—was able to increase the growth and the potassium content of the potatoes. Stoklasa[2] shows that potassium (in sealed test-tubes hung over growing plants) alters the rate of growth of the plants. We should like to add that we have been able to make this experiment, the validity of which we are able to confirm by having obtained corresponding results. Ried[3] shows that the presence of potassium and other salts in the vicinity of animals can have a far-reaching influence on their growth and above all on their reproduction. And there are numberless experiments which demonstrate the influence of irradiated and non-irradiated metals, in their effects at a distance upon the development of bacteria cultures. Only the Italian works of Rivera and Sempio[4] are cited here.

We are moving, as will be clear to the reader, no longer in the sphere of the substantial nourishment of plants, but in a sphere

---

[1] Dr. A. Heisler, *Aerztliche Rundschau*, 1932, No. 1.

[2] *Kosmos*, 1933, No. 12, Mittlg.

[3] Dr. O Ried, "Biologische Wirkung photechischer Substanzen", *Wiener Medizinische Wochenschrift*, 1931, No. 38.

[4] V. Rivera, *Sulla influenza biologica della radiazione penetrante*, Bologna Licinio Capelli-Editore, 1935. V. Rivera, *Ancora sull azione biologica dei metalli a distanza*, Roma, Scuola Tipografica, 1933. Caesare Sempio, *Rapporto fra effetti prodotti dai metalli posti a distanza, a contatto e in soluzione sullo sviluppo della thielaviopsis basicola*, Pavia, Premiata Tipografia, 1935.

which, for simplicity's sake, we refer to by the term "dynamic". The observation of these dynamic relationships of plant life as well as the substantial belongs to the biological part of this book. The choice of these highly diluted substances from the environment constitutes dynamic activity on the part of the plant.

The finest example of this activity is the Tillandsia usneoides or Spanish moss, a plant growing extensively in southern United States from Texas to Florida and in Central America. It is no saprophyte, but lives on trees and—what is still more remarkable—it is to be found sometimes on electric wires. Its mode of obtaining nourishment is something to give us pause for thought.

"This most remarkable of all the epiphytes which, in tropical and subtropical America, often completely enveils the trees, consists of threadlike shoots, often more than one metre long, with slender, grass-like leaves which—in the first stages of the plant's growth—are connected with the bark only by means of early-dried, weak roots. The fact that they remain fastened is due to the fact that the basal part of the axillaries winds around the stem portions. The sprouts are covered with scale-like hairs which are similar in form and arrangement to those of other bromeliaceae. The spread of the plant occurs less through seeds than vegetatively; sprouts are torn off and carried away by the wind or by birds which like to use them as material for nest building."[1]

The most interesting thing about this plant is the manner in which it obtains its nourishment. Since it is rootless, it can obtain no nutritive material from the surface on which it grows. Its intake of nourishment is accomplished by means of leaf and stem organs. Wherry and Buchanan[2] describe this as follows: "The Spanish moss is an epiphyte, requiring support by other plants and usually hanging from trees. It is not, as sometimes supposed, a parasite, since it draws no nutriment from the sap of its support, growing indeed even better on dead trees than on live ones, and moreover thriving on electric wires. The scales with which it is covered have often been interpreted as serving to retard transpiration, but Pessin has recently shown that the similar scales of the resurrection fern do not function in this way, and suggested that such scales on air plants more probably serve to hold water by capillarity while the

---

[1] Quoted from the well-known plant geographer Schimper.

[2] "Composition of the Ash of Spanish Moss", by Edgar T. Wherry and Ruth Buchanan, Bureau of Chemistry, U.S. Department of Agriculture, from *Ecology*, Vol. VII, No. 3, July 1926.

plant is absorbing therefrom the mineral constituents it requires. The mineral nutriment of the Spanish moss is evidently obtained then from whatever salts happen to be present in the rain which falls upon it, in water which drips on it from nearby trees, or in dust which is blown in by the wind. The composition of its ash should accordingly be of interest as affording an indication of the materials added by wind and rain to the soils of the regions where it grows. . . ."

The two authors then present certain mineral analyses of the tillandsia, followed by an analysis of the rain water. In these analyses two points are especially noteworthy: (1) The *high* percentage in the plants of iron (averaging 17 per cent), of silicic acid (averaging 36 per cent), and of phosphoric acid (averaging 1.85 per cent), and (2) the *low* percentage of iron, silicic acid, and phosphoric acid in the rain water of the regions where these plants grow (Fe : 1.65 per cent; SiO_2: 0.01 per cent; and no phosphoric acid at all). The authors therefore state:

". . . It is evident that the Spanish moss and other air plants exhibit selective absorption and accumulation of individual constituents to a marked degree. . . . Consideration of the composition of rain water shows that in general this plant *does not take up constituents in the proportions present in the water, but exerts a marked selective action.*"

In a later work[1] Wherry and Capen go further into the question of the content of mineral substances of the tillandsia:

". . . It seemed desirable to make further analyses in order to find out whether this plant would show any marked differences in mineral composition when growing on trees and when growing on electric wires close by, and also to compare it in this respect with other species of air plants growing in the same regions. . . ."

The authors continue with a series of analyses of different tillandsia plants grown under varied conditions. We cite No. 6 as a typical example, it being a tillandsia grown on a cypress tree, at Kissimmee, Florida, and No. 7, a tillandsia grown on an electric wire in the same place. The following tables represent analyses of the ash:

No.	Ash	Na_2O	K_2O	MgO	CaO	Fe_2O_3	SiO_2	P_2O_5	SO_3	Cl
6.	4.58	15.85	5.81	14.06	12.09	15.30	20.52	2.30	9.38	10.52
7.	5.15	12.96	7.75	8.67	13.28	18.60	28.76	2.90	3.27	4.87

[1] "Mineral Constituents of Spanish Moss and Ball Moss", by Edgar T. Wherry and Ruth G. Capen, Bureau of Chemistry and Soils, U.S. Department of Agriculture, from *Ecology*, Vol. IX, No. 4, October 1928.

SOIL FERTILITY, RENEWAL AND PRESERVATION

"Comparison of the two new analyses shows that the plant growing on wire is notably higher than that growing on the trees in ferric oxide and silica, but lower in soda, magnesia, sulphur and chlorine. The significance of these differences will be taken up after the results of analyses of another air plant are presented."

There are other plants which have the capacity for growing on electric wires and nourishing themselves. To these belongs the ball moss (Tillandsia recurvata). The authors also give a series of analyses of these, of which we here present two. The plants grew in the same place as the others, No. 2 on an elm, No. 3 on an electric wire.

Analyses of the ash of ball moss:

No.	Ash	Na_2O	K_2O	MgO	CaO	Fe_2O_3	SiO_2	P_2O_5	SO_3	Cl
2.	5.52	10.72	10.90	6.97	17.19	15.74	25.07	3.78	7.20	4.96
3.	4.27	11.19	5.55	10.27	13.89	15.80	31.72	2.50	5.76	5.87

2. Plant from tree, Kissimmee, Florida.
3. Plant from wires, same locality.

Here, too, there are clear differences, although partly in another direction from those of the above-cited Tillandsia usneoides. From still another analyses it becomes plain that the further we go inland, in a dry climate, the higher is the content of silicic acid. It is universally observed that the content of mineral substances is in general a very fluctuating one. The two plants tested had grown under identical conditions, except for their bearer, and were touched by the same rain, dust, etc. The authors of the report conclude:

"... The fact that in the latter case the rain falling on the plants could have had no previous contact with any source of mineral matter other than particles of spray or dust carried high in the atmosphere, *brings out in a striking manner the ability of air plants to extract relatively large quantities of inorganic constituents from exceedingly dilute solutions.*"

This quotation directs our attention in a positive way to a sphere of plant physiology which has heretofore been observed, but which has so far called forth but little research: the absorption from the air of highly diluted substances of which only very slight traces exist. Under the conditions prevalent in Florida, the Spanish moss, with its rich mineral content, would be an extremely valuable help if used in making compost for fertilizing the soil.

When the author some years ago called attention to such facts, he drew upon himself the criticisms of a number of scientists. His laboratory researches presented at that time—which, let it be said, had been checked and controlled from many angles—were regarded with amusement. In view of the prejudices of certain circles with regard to this problem, there was nothing to be done except to call attention to certain phenomena evident in nature. The prejudices aroused by observations and research of this sort are all the more difficult to understand, in view of the existence of a whole series of examples of this "selective absorption of the highest dilutions", as cited on page 118. But now it has already become "commonsense" to speak of the importance and value of traces of iron, copper, zinc, manganese, for plant health. As a matter of fact the circumstances are not always to be seen as clearly as in the case of air plants, because we are usually dealing with plants which grow in ordinary earth.

Apparently the farmer does not directly influence these finer processes—and yet the health and the intensity of growth depend on them as on soil cultivation and fertilizing. At first a state of equilibrium arises in nature. This is disturbed by men or by outer circumstances, for instance, by intensive fertilizing with unbalanced, not "buffered",[1] minerals. But Nature acts to help herself. In the selective capacities of the plants, Nature herself creates "one-sidedness" or "specialization". But these conditions of one-sidedness only arise when it is their function to heal the effects of another "one-sidedness" of nature.

We see in this no mere chance, but rather a wise foresight on the part of nature. We must distinguish in this connection the bulky, heavily fed plants which grow only on good soil, with good intensive fertilizing; these constitute the greater part of our cultivated plants. They collapse with the appearance of any phenomenon of deficiency. These plants must be distinguished from the dynamically active plants which, on their side, help to correct the unbalanced state of nature. We must study these plants very exactly in their relationships to nature, and then we shall discover one of the important secrets of biological phenomena. We shall present a number of examples in illustration of this principle: *tobacco* is rich in potassium when it grows in a soil poor in potassium, and vice versa. The wood and bark of *oak trees* are especially rich in calcium (up to 60 or more per cent of the ash is CaO). Moreover, they can grow in sand—that is, in a soil poor in calcium

---

[1] "Buffered"=harmonized through the organic colloidal condition of the soil.

SOIL FERTILITY, RENEWAL AND PRESERVATION

—and despite this are able to accumulate calcium. And their iron content may reach 60 per cent. *Buckwheat* is a typical sand-silica plant, and distinguishes itself by its accumulation of calcium. Here we see dynamic-selective conditions. The *common broom* (Cytisus [Sarothamnus] scoparius) is an especially remarkable plant. It is conspicuously rich in calcium in its stamens and leaves (25.03 per cent CaO where the soil has only 0.35 per cent CaO).[1] Besides this, it throws off calcium from the roots, depositing it in encircling rings, thus providing the soil with calcium. Especial attention is called to this phenomenon which occurs in a predominantly sand plant. These qualities make this plant particularly suited to help in preparing dry waste lands and the sandy prairie, and, by an improvement and enrichment of their calcium content, to win them back to a state of cultivation.

On the lovely English lawn—constantly increasing in acidity and decreasing in calcium—daisies may appear. They are remarkably rich in calcium. They signify, at the same time, the fact that the soil has passed a certain "acid limit", and it is their task to garner the substance it needs. In all such cases we may ask: Whence comes this substance, how does the plant take it up? The so-called "weeds" become enormously significant in this connection. They are only weeds from the human, utilitarian point of view. At most we might call them "misplaced" good plants, good plants growing in the wrong place, as it were. But a more exact study shows that they are very much in the right place. They are the most accurate specialists. They have adjusted themselves to a very definite degree of acidity, so that their mere presence gives us an exact indication of the acid or alkaline state of the soil. They are the "stop lights" or "warning signals" of soil life. Let us cite a few more of their specific qualities, in order that we may then draw a conclusion from all this.

Relationship to magnesium: Fagus silvatica becomes rich in magnesium with increasing age, with 12.4 per cent in a 10 year old, up to 19.5 per cent in a 220 year old tree (shown in the ash). Oak and Picea excelsa, on the contrary, show a decreasing magnesium content with increasing age (the oak, for example, shows 13.4 per cent at 15 years and 2.35 per cent at 345 years.) Barks of trees are in general poor in magnesium; birch, on the contrary, is conspicuously rich in it, showing often as much as 14 per cent. Normally its leaf ash contains from 3 to 8 per cent.

[1] Gustav Hegi, *Alpine Flora*, quoted from the German original, Vol. IV, p. 1185.

THE DYNAMIC ACTIVITY OF PLANT LIFE

Specialists in accumulating magnesium are:

	(MgO content in percentages)		(MgO content in percentages)
Prunus	12.3	Solanum tuberosum	28.5
Acer campestris	10.5	Stellaria media	21.8
Erica carnea	15.5	Ilex Aquifolium	20.6
Betula alba	15.3	Herniaria glabra	18.9
Scrophularia nodosa	15.6	Spiraea ulmaria	18.0
Beta vulgaris	25.9		

Fat-accumulating seed pods contain in general more MgO.

Grain varieties	11.0	Linum	14.2
Maize (corn)	15.5	Gossypium	16.6
Fagopyrum	12.4	Cannabis	5.7
Pinus	7.9	Theobroma	11.0
Phaseolus	7.6	Cocos	9.4
Castanea	7.47	Aleurites	15.1
Quercus	5.2	Juglans	13.0
Brassica rapa	11.8	Abies	16.7
Brassica Napus (rape)	13.4	Amygdalus	17.7
	Papaver 9.4		

Wood ash in general, 5 per cent to 10 per cent MgO.

Worthy of special note are: Rubus fruticosus, 15.81 per cent; Betula, up to 18 per cent; Quercus, 15 per cent to 23 per cent; Larix, up to 24.5 per cent. In all these, there are also rhythmical variations in the course of the year.

The farmer may now ask: "Well, what significance has all this for me?" The significance lies in the fact that many of these plants live in or near his own cultivated areas, and also, that at certain times they fertilize the soil by dying off, by dropping their leaves, etc. More important, however, is the fact that they furnish, in organic, available form, just that finely diluted substance which nature needs for healing or stimulating its life processes. The remarkable thing is that many of these plants are just "specialists" in producing the substance which the soil lacks, thus contributing to soil improvement. They accumulate, as we have seen, these substances from states of high dilution and then, by condensing them into more concentrated form, carry them into the soil.

An imitation of this process can be accomplished in a practical way by making compost of *everything*. The greater the variety of plants used in making compost the richer and more useful it is in its nutritive potentialities. The compost made of weeds or of medicinal herbs (both are, in this respect, closely related) can carry

SOIL FERTILITY, RENEWAL AND PRESERVATION

to the soil just what can be brought to it in no other way, in a form organically in harmony with it. Since, to a large degree, we have to do with dynamic effects, the most minute quantities of this substance, or that, may often be sufficient.

It may be of interest and value to mention a few more special activities of plants. For instance *henbane* and *thorn apple* (Datura stramonium) like to grow in a soil rich in calcium. Their ash content is relatively poor in calcium, while on the other hand they are both "specialists" in the formation of phosphoric acid; henbane with 44.7 per cent and thorn apple with 34.7 per cent. The common *foxglove* (Digitalis purpurea) likes iron, and also calcium and silicic acid; besides this it stores up manganese. The *dandelion* is fond of calcium and silica. Its ash residue is especially rich in these. It grows luxuriously in countries where alfalfa is cultivated. The *Robinia pseudacacia* or yellow locust—the significance of which as a basis for afforestation we treated in Chapter IX—likes sandy soil and it accumulates calcium up to 75 per cent. Its leaves, in a mixed forest, are the best fertilizing agent against a deficiency of calcium. In addition it gives nitrogen to the soil.

An interesting plant is the *sugar beet*, growing in the vicinity of the sea. It is a small "pharmacy" with up to 56 per cent of its content composed of sodium, lithium, manganese, titanium, vanadium, strontium, caesium, copper, rubidium. Where fertilizing is one-sided, this "many-sidedness" of the sugar beet disappears. How could the sugar beet feel comfortable under these conditions? Wild sugar beets grow at the seashore, along deeply indented bays and fiords. There they are fertilized by nature with a "compost" of seaweed. This sea is again itself a "pharmacy" and contains many of the ingredients of which the sugar beet is fond. Should we not make the sugar beet "happy" by giving it a few small doses of a compost made of seaweed, or its ash, served up as a "dressing", in order to remind it of its primeval origin? Boron, for example, is also present in such a seaweed compost. In this way we should help the sugar beet far more, and in a more natural fashion, than by any one-sided artificial treatment.

The *Chrysanthemum segetum*, a weed, prefers loamy and clay soils, deficient in calcium. The plant itself is rich in calcium, phosphoric acid and magnesium. This should be turned to advantage.

Sheep's sorrel (Rumex Acetosella) is to be found on sour parcels of land. It tells us that the pasture soil is becoming too matted and closed up. Its ash is rich in calcium, phosphoric acid, magnesium and silicic acid. Into the compost pile with it! Let it fertilize the same pasture again—after harrowing.

Cochlearia Armoracia, the horse radish, is a provider of calcium (11.9 per cent Ca), of phosphoric acid (13 per cent P_2O_5) and of sulphur (18.64 per cent SO_2). Potatoes do well in its vicinity, too.

German camomile is rich in salts in general, especially potassium (45 per cent) and calcium (23 per cent). Its richness in sulphur compounds is conspicuous.

Cacti contain much calcium. There are varieties which accumulate lime up to 80 per cent of their dry substance.

Equisetum varieties are strong gatherers of silicic acid, like grasses which have leaves rough to the touch when stroked downwards from tip to base.

Yarrow is rich in potassium, calcium and silicic acid.

The *stinging nettle* (Urtica dioica) is rich in calcium (36.4 per cent) and silicic acid.

Urtica urens shows incrustations of silicic acid on its stinging prickles.

The *onion* is a "specialist" in silicic acid, and also in calcium in its leaves.

The above is offered only as a small segment of this field of study. Much in this field is to-day still uninvestigated and untouched. It is essential to direct biological study toward this sphere which is very important for the practical man. *But of one thing we are sure:* the more varied the compost, the more certain also are its dynamic effects. Some of these plants have importance in directing the fermentation of organic matter.

In Chapters VIII and IX we have already treated the significance of a mixed culture. The presence of weeds must be considered as a dynamic factor. The *orache* (Atriplex hortensis), or pig weed, prefers to grow on humus soil which has been treated with good organic fertilizer. This plant has an especial affinity for the potato. While the potato otherwise chokes root herbs, the orache likes to grow next to it and check its growth. It can even happen that soils exhausted by potatoes indicate this fact by showing an increase in the growth of the orache, just as soils tired out from hoed crops in general frequently show this through an increase of *black nightshade* (Solanum nigrum); tired soils which become crusty because of a too constant planting of grains show it by an increase of *German camomile*, and soils overfertilized with potassium show it by the spread of *hedge mustard*.

The mutual influence of plants on one another belongs in any case to a borderline sphere of dynamic and biological effects. If *wheat* and *poppies* are grown together the poppies check the development of the wheat, and thus lower the yield of wheat kernels, while, on the other hand, the *cornflower* is less harmful. The author has conducted experiments for some years in regard to plant symbioses. In these it has been found that the *German camomile*, for example, can be both good and harmful. If there is only an occasional camomile plant growing in the middle or on the border of the experimental field, planted with grain, then the German camomile fosters its growth. If, however, it is planted thickly, for example in rows through the grain or along the border, then it has an inhibiting effect. Here we are dealing with a typical dynamic effect.[1] In further experiments it was possible to observe other stimulating and inhibiting effects. Some of these tests were arranged as plant experiments in the open field, others as seed baths with extracts of the plants in question (one half hour bath and then sowed and grown normally), others as germination tests after the seed bath, others as tests of growth in a seed bath of corresponding dilution. In practice, we can make use of the knowledge of the effects of the so-called "border plants" sown along the edge of the bed or field of cultivated plants. The indications and impulse for this investigation came from Dr. Steiner and have proved to be of great value.

Tested by the most varied methods, the following have shown themselves to be especially "helpful": *dead nettle* (Lamium) and *esparcet*. Another good border plant is *valerian*, and here and there *yarrow*. The following are harmful: German camomile in too great quantity, buckwheat, hemp, cornflowers in too great profusion, poppy.

Lippert has called attention through his investigations to the significance of the stinging nettle. Planted in rows between medicinal herbs, it raised the content of ethereal oils in these plants in comparison with the control plots. It was further proved that the pressed-out juices of plants, which are grown alongside stinging nettles, mould and putrefy less easily.[2]

From the literature[2] on the subject, the following is also familiar. Rye is hostile to wheat growth, inhibiting germination and growth

---

[1] Cf. Prof. Boas, *Ueber Hefewuchsstoffe*. German camomile, in a dilution of 1 : 8,000,000 stimulates the growth of yeast. Other substances stimulating yeast growth are effective in dilutions of 1 : 4,000,000.

[2] Dr. J. Kuhn, *Mitteilungen der Deutschen Landwirtschaftsgesellschaft*, Vol. I, 1932.

THE DYNAMIC ACTIVITY OF PLANT LIFE

of the field poppy, among other plants. Where there is a strong growth of couch grass (Triticum repens), rye planted twice in succession can eliminate this troublesome weed. Poppy and larkspur show an affinity for winter wheat, but not for barley. Seeds of these weeds lying in the ground awaken to activity when, in the rotation of crops, winter wheat reaches its turn.

Hedge mustard and wild mustard are fond of oat fields, and in this case they signify no inhibiting symbiosis. On the contrary, they both work inhibitingly in the case of rape. Especially harmful for turnips are the hedge mustard and the knotweed. Red clover and plantain, lucerne and dandelion like to grow together. The effect of rye on wild pansy is extraordinary. While the latter in general only germinate from 20 to 30 per cent, they germinate up to 100 per cent in a rye field.

If these facts are applied in the right way, the planting of border and "protective" plants can be developed and can be of great assistance in combating the effects of present day one-sided field cultivation, and be of special importance in the cultivation of gardens and the growing of medicinal plants.

This chapter is inserted in our consideration of this whole subject in order to illustrate further what may be understood by the expression *"the bio-dynamic principle in nature"* and to show how we may learn to make use of it.

CHAPTER TWELVE

Scientific Tests

There are two objections advanced by opponents of the bio-dynamic method of agriculture. One of these has to do with the effect of the preparations. The special handling of the manure and compost—so we hear and read—is described as exemplary, and recently it has even been referred to as something obviously correct. But considerable doubt is expressed in regard to the effectiveness of the preparations, or that they have any special effect. The other objection has to do with the long-term effect of the bio-dynamic method. If we were to work "in that way" we would be "robbing" the soil of its nutritive materials, so the objectors say. We have already discussed the second question in Chapter Two. In Chapter Thirteen we shall give practical agricultural results which show this fear to be unfounded.

A series of experiments has been carried out by the author and his associates in the chemical-biological research laboratory at the Goetheanum. The simplest experimental procedure is the following: Seeds, chopped potato, etc., are bathed for one half hour to an hour in a thin dilution of preparations 500, 501, 502, 503, 504, 505, 506, 507 (all these are made from directions given by Dr. Rudolf Steiner). Above all, the development of the roots is observed. A variation of this method is to let the plants grow directly in the corresponding dilute solution. In every experiment performed in this way the result has shown that preparation 500 has a particularly stimulative effect on the root growth, and causes the development of numerous fibrous roots. Preparation 501 increases the assimilative activity of the plant, preparations 502–507 strengthen the plants in general in their growth. In this connection 504 has an especial influence on the quality of the flavour, the others more on the productivity of mass. The exact results of the experiments are given in the monthly periodical *Demeter*.[1]

[1] 1931, No. 7, July, E. Pfeiffer: "Pflanzenversuche zur biologisch-dynamischen Wirtschaftsweise". 1935, No. 7, July, E. Pfeiffer, E. Künzel, E. Sabarth: "Versuche zur Wirkung der Präparate 500, 501, sowie 502-507".

SCIENTIFIC TESTS

We present here two experiments illustrative of these results:

Lupins are germinated in a water culture with the addition of the various solutions, and planted out in soil. This is followed by irrigating all the plants with the same tap water. After 10 days the length and weight of the plants are measured. We have, as before, controls germinated in tap water alone, other plants germinated in baths of 500, and others with 501 in a 0.005 per cent dilution, and another group of plants germinated in the 500 solution which have their leaves sprayed with 501—seven days after planting out—in the dilution used in ordinary practice (shown in the table as 500 plus 501).

Mean values for 20 plants in each case:	control	500	500+501	501
Root length after 19 days, in cm.	11.08	13.74	15.16	14.5
Root weight after 19 days, in grm.	0.363	0.443	0.726	0.452
Sprout length after 19 days, in cm.	9.5	10.9	13.0	12.5
Sprout weight after 19 days, in grm.	2.071	2.093	2.426	2.42

Result: The plants treated with the preparations are superior to the control. The arrangement used in general practice: preparation 500 for germination, with preparation 501 sprayed on the leaves after transplanting; results, maximum.

An experiment with wheat. Germination and growth in sand, containing humus. Growth measured after twelve days. The soil of the control was moistened with tap water, the others had respectively 500, 501, and 500 plus 501, in the dilutions previously given. Preparation 500 was applied before sowing, preparation 501 was sprayed on the leaves.

Mean values for 30 plants in each case:

	control	500	501	500+501
Sprout length in cm. first leaf	11.8	10.7	10.5	11.6
Sprout length in cm. second leaf	5.6	4.7	6.1	6.5
Root length in cm.	17.2	22.7	20.2	22.0

The result corresponds to previous experience.

To the figures on the experiments with Lupinus albus, let us add some photographs of typical test plants. In these, the forms of the roots are especially to be noted.

The wrong, too early use of preparation 501, especially in number 6, leads to atrophy and weak development of the root.

Especial attention was given to the study of the stimulative effects of preparation 500 used together with light applications of bio-dynamically treated compost.

The value of nitrogen bacteria for the legumes and for general fertility of the soil is well known. Hence, it was important to determine whether the formation of nodules on the roots of the legumes is influenced by this procedure. For this experiment we used field soil which had been fertilized bio-dynamically for a number of years and other soil from ground of the same kind, which had, however, been treated for a number of years with so-called "complete fertilizing" (organic plus mineral fertilizing). Equal amounts of these two soils were separately mixed with pure sand. The soil was a heavy clay and had to be made somewhat looser by an addition of sand.

Then the mixtures were put into flower pots and at the same time a planting of kidney beans (Phaseolus vulgaris) was sowed in both. The seed was carefully selected and equal in weight. The experiment, carried out in the greenhouse, was concluded after the beginning of the fruit formation. The plants were carefully freed of the soil, washed, and the bacteria nodules carefully separated from the roots. The nodules of from 2 to 4 mm. size were counted, and then the total weight of all the nodules (those smaller than 2 mm. as well as those larger) was determined.

Result: weight of the root nodules, reckoned on the basis of 100 plants:

Bio-dynamic	16.2 grm.
So-called "complete fertilizing"	9.5 grm.

It is clear from this that the nodule formation is substantially benefited through bio-dynamic fertilizing. If we consider the high economic value of legumes in agriculture, especially their function in the crop rotation, a wide perspective opens up from this experiment. The fixation of atmospheric nitrogen (which cannot be taken up directly by the plants) into organic nitrogen compounds is accomplished by the root nodules of the legumes. The furthering of this activity means enriching the soil in nitrogen compounds which may be taken up by the plants. We see here one of the chief merits of the bio-dynamic method of agriculture: that it furthers the natural activity of the soil life in a way that has not been observed when other fertilizers are used.

The significance of the earthworm for humus formation in the soil has already been discussed in Chapter Five. It has come to light in practice that the bio-dynamic preparations improve especially the general conditions of the earthworm's environment. Hence the earthworm migrates towards these more favourable conditions. This can be proved by simple laboratory tests.[1]

---

[1] J. v. Grone-Gueltzow: "Wie verhält sich der Regenwurm zu biologisch-gedüngtem Bodem". *Gäa-Sophia. Bd. IV: Landwirtschaft, 1919.* Dornach.

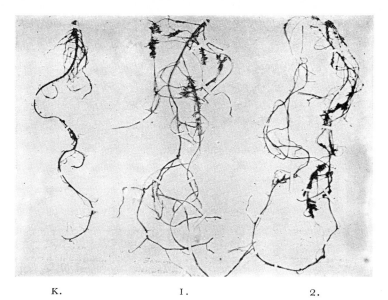

K.   1.   2.

PLATE 6. K. Control-germinated in tap water. 1. Pre-treatment with 500. 2. Pre-treatment with 500 and again before transplanting. (Lupin experiments)

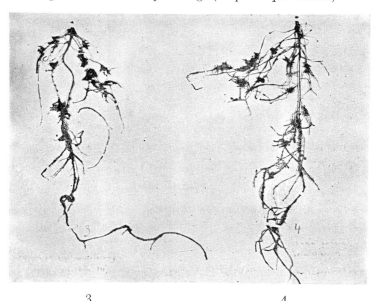

3   4

PLATE 7. 3. Germinated in 500, 501 sprayed on leaves (note in 2-4 the strong development of the root hairs and the bacteria nodules). 4. Germinated in 500, 500 at transplanting, later 501 sprayed on leaves

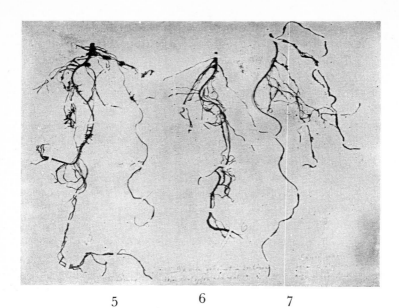

5   6   7

PLATE 8. 5. Germinated in 501, long, thin root. 6. Germinated in 501, 501 again at transplanting. 7. Germinated in 501, 501 again at transplanting, 501 sprayed on leaves (wrong treatment in cases 6 and 7)

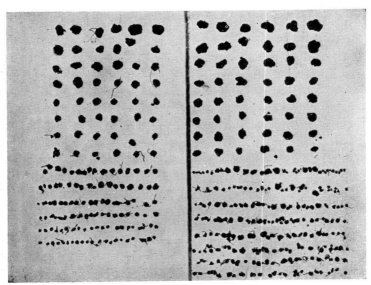

PLATE 9. Nitrogen bacteria nodules
Left: usual treatment. Right: bio-dynamic treatment

SCIENTIFIC TESTS

We now come to a series of experiments which were made in the Research Laboratory at Dornach. The procedure was as follows: a wooden box was divided into four equal compartments filled respectively with the same earth variously treated. One portion had been watered with a customary dilution of fertilizers containing potassium, nitrogen and phosphoric acid, the second with a dilution of urine. In the third division, earth was deposited which had been fertilized with our biological preparations, in the fourth was the same earth unfertilized, as a control. Needless to say, the same compost earth was the basis for all four tests, and all four divisions were kept equally moist during the experiment. Into each part we placed the same number of earthworms, and after an interval of several days had elapsed a check was made to discover the location of these worms.

There were small openings between the individual divisions, so that the worms could crawl through them out of the soil in which they had first been placed, in order, as we have the right to assume, to choose the soil which suited them best.

Here are the figures from three of a series of such experiments:

Worms set out		Found after 1 day	Increase or decrease
Artificial fertilizer	11	9	—2 = —18 per cent
Urine	11	8	—3 = —27 ,,
Bio-dynamic	11	13	+2 = +18 ,,
Control	11	12	+1 = + 9 ,,
		Losses: 2	

Worms set out		Found after 3 days	Increase or decrease
Artificial fertilizer	10	5	—5 = —50 per cent
Urine	10	5	—5 = —50 ,,
Bio-dynamic	10	24	+14 = +140 ,,
Control	10	5	—5 = —50 ,,
		Losses: 1	

Worms set out		Found after 4 days	Increase or decrease
Artificial fertilizer	10	8	—2 = —20 per cent
Urine	10	2	—8 = —80 ,,
Bio-dynamic	10	22	+12 = +120 ,,
Control	10	6	—4 = + 40 ,,
		Losses: 2	

We see that the result was surprisingly in favour of the bio-dynamically fertilized soil. The urine and the artificial fertilizer were readily forsaken and the bio-dynamically fertilized soil was much preferred by these creatures, which, boring their way through the ground, taste with their soft, slimy skins.

Above and beyond the bare figures, which certainly speak a plain language, we were also able to observe how the "bio-dynamics" swiftly disappeared into their soil, in especial contrast with the worms put on urine-saturated soil; these remained for quite a while lying inert on the surface.

It is especially noteworthy that the soil itself in which twenty-four rain worms had gathered had an obviously different and greatly improved structure after three days had passed. It is also worth noting that preparation 507, a specially prepared extract of valerian, has an attraction for earthworms—not only for cats! As we have seen, the regular bio-dynamic practice is to give a fine spray of this preparation to the completed compost or manure heap.

In a test of the effect of this 507 on earthworms, the following results were obtained after five to eight days (in four communicating compartments, half of which were filled with earth, others with earth plus 507):

	Number of earthworms	
	At the start	At the end of the experiment
Experiment one:		
Earth	10	9
Earth plus 507	10	18
Experiment two:		
Earth	10	8
Earth plus 507	10	19 (the additional worms moved in from outside)

One can gather from this that the spraying of the heaps with preparation 507 has an effect of attracting the micro-organisms, etc., of the soil, and brings them in swiftly to start their activity.

Another experiment was carried out according to the classical method of Dr. A. Wigmann and L. Polstorff mentioned in *The Inorganic Constituents of the Plant*. This, it may be remembered, was honoured by the University of Göttingen in 1842. In its time this method served for a determination of the nourishment and the mineral salt requirements of plants, out of which then came consequences of such far-reaching importance for mineral fertilizing. The method rests on giving a plant the possibility of letting its roots grow simultaneously in variously fertilized soil and then determining in which direction the preferred growth goes.

PLATE 10. Photo of worms
A spadeful of three-months-old bio-dynamic compost.
Note the great quantity of earthworms

PLATE 11. Root growth. This photograph shows chrysanthemums which have grown down into the biodynamically fertilized earth with the chief mass of their roots on the right side

SCIENTIFIC TESTS

As this was applied to our problem, flower pots were filled with ordinary earth. The bottoms were removed from these pots and they were placed on the edge of partitions dividing a box into two compartments, so that they occupied an equal space over each compartment. In one compartment earth, fertilized with ordinary compost, was placed; in the other was the same earth fertilized with the same compost, but the latter had been bio-dynamically prepared. When plants were set out in the pots and permitted to grow, the roots grew into the one or the other half of the box.

Thus the plant itself demonstrated its ability to find the conditions of life most satisfactory to it; and for practical purposes the

FIGURE 6. MODUS OPERANDI OF EXPERIMENT

last and surest criterion in the investigation of a method of agriculture must always be indicated by the way the plants themselves react.

The experimental arrangement just described was also used to show the individual effects of the preparations. Test plots were divided in two sections by a separating glass partition. The sections were covered with untreated compost; but, at a lower level, on one side of the partition bio-dynamic preparations were inserted, while on the other side of the partition no preparation was used. The plants grew first in neutral earth, then they continued their growth, their roots working down into the variously treated soil, and developing variously in accordance with this. With this experimental arrangement we were able also to confirm the effect of the preparations.

SOIL FERTILITY, RENEWAL AND PRESERVATION

Rate of growth of soya-bean roots in bio-dynamically treated earth and in the same soil, untreated:

Preparation Number	Average percentage of increase as compared with control	
	Length	Weight
502	+ 4.5	+ 1.6
503	+12.8	+ 6.9
504	— 5.2	—13.8
505	+ 4.5	+14.2
506	+ 4.5	+22.7
507	+ 5.1	+10.8
502–507	+ 5.8	+20.5

This table shows a whole series of interesting results. It demonstrates one thing plainly, that definite effects are produced by the preparations. In one case, that of the 504, we find that the quantitative effects are inferior to the control. But in connection with this it must be noted that other experiments have proved that this preparation brings about an intensification of qualitative characteristics, such as aroma, etc.

The table also shows that the effects on the growth of the root and on the weight are quite varied. It is clear that this is no ordinary growth stimulus, but that the whole structure of the root also, its density and water content are altered.

Seed-bath tests with the bio-dynamic preparations have shown similar differences. In these the seeds to be used in the experiment were placed for twenty, thirty or sixty minutes in a highly diluted solution (for example 0.005 per cent) of the preparation in question, dried and then immediately sowed. It was found that when the seeds were permitted to lie for several days after the bath the beneficial effect was lost. Such seed-bath experiments have been made for a number of years in the author's laboratory and in other places, and are sufficiently extensive to permit a definite judgment of the effect of the preparations. Since in these tests only small quantities of material in high dilution are used, and only for brief periods of time—after which the seeds are planted in the same, ordinary soil and grown under normal conditions of temperature and moisture—*it is plain that we are here dealing with a purely dynamic effect*.

Average value for twenty plants in each case: (a) weight of radishes including roots (b) weight of leaf mass. The experiment is concluded when the radishes are "ripe".

SCIENTIFIC TESTS

	(a) in grm.	(b) in grm.
Control	2570	2530
Seed bath with preparation 500	2775	2655
Seed bath with dilution of prepared manure	3010	2300
Seed bath with preparation 500 and 502–7	4600	4090

Such seed-bath experiments show very clearly the effect of the preparations in stimulating growth. Here we are confined to a presentation of principles. The complete results of the experiments in this field will be published at some future time.

Average value of 10 maize plants in each case:

	Height of plant in cm.	Weight of roots in grm.	of leaves and stems in grm.	of ears in grm.
Control	181	50	2181	727.5
Seed bath with 500 and 502–7	185	65	2250	800

It must be emphatically stated that the plants used in making the bio-dynamic preparations would not produce the same effect by themselves *before* being put through the bio-dynamic process of preparation which they produce *after* this treatment. Comparisons have been made of seed baths containing extracts of the selected fresh plants with seed baths containing extracts of the same plant materials after they have been put through the bio-dynamic process. The variations of growth and root development on the tested plants have been carefully observed and noted. In addition, a new factor of "usability" has been set up: a compound of *flavour* and *uniformity*, size and shape. From the results, we see first that the effects of the preparations are very varied, and secondly that this variation shows itself in the quantitative and qualitative aspects of the plant treated with them. *And it is important to note the fact that the very brief seed-bath treatment influenced the later growth of the plant.*

The forms of the tested radishes and control plants show very characteristic differences.

Control:	taste—juicy, bitter	Usability	27.5	per cent
Urtica dioica (stinging nettle):	taste—insipid	,,	16.6	,,
Preparation 504, made from Urtica:	taste—juicy, stimulating	,,	26.9	,,
Oak bark:	taste—woody	,,	34.37	,,
Preparation 505, made from oak bark:	taste—dry, sweet	,,	42.85	,,
Taraxacum officinale (dandelion):	taste—juicy, delicate	,,	13.63	,,

SOIL FERTILITY, RENEWAL AND PRESERVATION

Preparation 506, made
from Tarax. off.:   taste—stimulating
                      aromatic   Usability 63.63 per cent

Another result of the experiment was that it showed that the yield in terms of mass and the yield in terms of quality do not always coincide.

Experiments with the bio-dynamic preparations, carried out by the Greek Government Office for the Protection of Tobacco in Salonika, show interesting results both quantitatively and qualitatively as compared with artificial fertilizers and untreated soil. The preparations—numbers 502, 503, 505, and 507—instead of being used in compost and manure in the usual way, were stirred in water and sprayed directly into the soil (as directed for preparation 500). Preparations 500 and 501 were employed in the usual way. The tobacco yield, without other manuring treatment, was superior to that obtained with a dressing of "Ammo-phos" (14-28-12), and only slightly inferior to the yield of a nitrogen-phosphoric acid-potash fertilizer (6-8-8). Growth was rapid and regular, and leaves were erect and of deep colour.

Experiment 1

Plot number	Manuring treatment	Yield in "okes" (per 500 sq. metres)
5	Nit., phos. ac., pot. (6-8-8)	$47\frac{1}{2}$
3	Preps. 500, 502, 503, 505	45
6	Prep. 500	44
1	"Ammo-phos". (14-28-12)	42
2	Control (no fertilizer)	35
4	Control (no fertilizer)	33

Experiment 2

A further experiment with bio-dynamic preparations yielded the following results:

Plot number		Yield in "okes" (per 2500 sq. metres)
3	Preps. 500, 502, 507	190
1	Prep. 500	132
2	Preps. 500, 503, 505 (damaged by hail)	122
4	Control (no fertilizer)	121

A preliminary report on the quality of the tobacco, received before the process of fermentation was completed, gave the following results:

(a) Quality as regards colour:

Plot number		
1	Prep. 500	1st
3	Preps. 500, 502, 507	2nd
2	Preps. 500, 503, 505	3rd
4	Control	4th

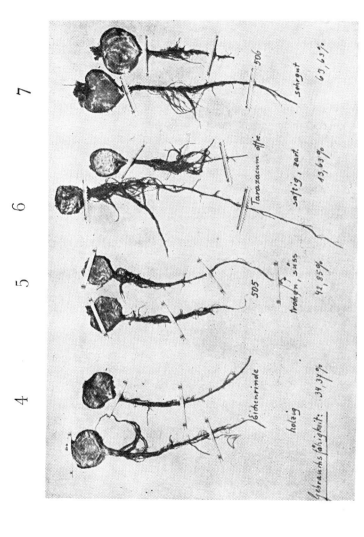

PLATE 12A. Radishes treated with oak bark, preparation 505, dandelion, preparation 506

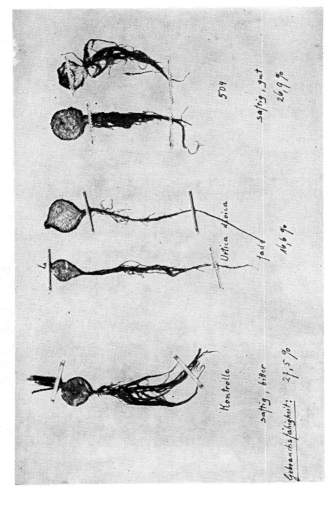

PLATE 12B. Control radish, radishes treated with stinging nettle and with preparation 504

SCIENTIFIC TESTS

(b) Quality as regards aroma, texture, etc.:

Plot number		
3	Preps. 500, 502, 507	1st
1	Prep. 500	2nd
2	Preps. 500, 503, 505	3rd
4	Control	4th

We quote in full the conclusion of the report:

(1) The bio-dynamic preparations have given satisfactory results as regards quality.

(2) By the use of the preparations the tobacco plants developed more rapidly and reached maturity some days earlier.

(3) Plot No. 3 (500, 502, 507) gave the best combined results. Plot No. 2 (500, 503, 505) came second. This plot, notwithstanding its apparently small yield, competed effectively with the previous plot as regards both quality and quantity, for we believe that if this plot had not suffered from severe hail which destroyed 40 per cent of it, it would have been, perhaps, the best of all. Third place was taken by plot No. 1 (500) and lastly came the control plot.

Contrary to our first opinion, we believe, or rather we are convinced: (1) That these preparations, supplemented rationally and utilized with care and attention, can become beneficial in every respect to the cultivation of tobacco. (2) That it is indispensable to continue the experiments with all the preparations for at least 2 more years, in order to enable their inventor and ourselves to draw more precise and positive conclusions for the benefit of the Greek tobacco industry.

In regard to the carrying out of such experiments, it must be added that a very rich, heavy soil produces less strong variation in the results of the experiments than when they are made on a poor or light soil. The various seasons of the year have to be taken into consideration, because all conditions that cause luxuriant growth lessen the experimental variations.

From these experiments it is evident that a very stimulating effect on growth comes from the preparations. On that account the small amounts used are sufficient; for example, we spray 40 grammes of preparation 500 diluted in 10 litres of water on one acre, or we insert from one to two grammes of each of preparations 502–507 into from one to three cubic yards of manure.

We know from the study of vitamins, hormones and ferments that effects of increase can be produced by small doses of certain biologically active substances. But if we attempt to apply this knowledge in an agriculturally practical way, we very often encounter the incredulous amazement of the so-called practical man; it is as though there could be no possible progress in this sphere! Naturally, we have to begin with the proper basis—that is, by providing for organic fertilizing and treatment of the soil—for it is this soil which is to be stimulated to an increased productivity by the use of the knowledge given above.

The effect of bio-dynamic treatment shows itself not only in changes in manure and soil; it extends to the quality of the food and fodder grown on it. In the case of wheat, an increase was first of all observed in its germinating capacity and germinating energy.

Result of six germinating tests with 350 grains of wheat:

Kinds of wheat	Percentage of wheat germinated in distilled water	Percentage of weight increase of seedling after 14 days (measuring only green plants)
Wheat bio-dynamically grown	99.3	98
Wheat (same species) grown on identical soil but according to previous methods (i.e. with the fertilizing materials but without bio-dynamic treatment)	92.3	86

With wheat, another difference was also apparent, in the gluten content and the gluten quality. With Dutch Juliana wheat, which was grown in neighbouring test fields, the following was demonstrated:

	per cent
Gluten *content* of the Juliana wheat first year, before the bio-dynamic treatment (original seed)	30.2
Gluten *quality:* elastic, well formed, of regular structure and normal rising energy	
Gluten *content* of the wheat second year, after growing with bio-dynamic treatment	41.6
Gluten *quality:* elastic, well formed, excellent rising capacity	
Gluten *content* of the wheat second year, on same kind of soil but with mineral fertilizer	28.8
Gluten *quality:* medium elasticity, irregular structure, medium rising capacity	

The growth of plants is outwardly determined by the kind of soil, nutrition, humus, climate, light, moisture or aridity, and other environmental factors. The soil-fertilizing air and moisture provide the plant with substances necessary for the building up of its main bulk.

In recent years a special study has been made of another group of substances, present in plant life in very minute quantities, but the bearers, it would seem, of the true dynamic forces. These substances are known as hormones (growth hormones in particu-

SCIENTIFIC TESTS

lar), ferments, vitamins, etc. They are not nutrition in the sense of minerals, proteins and starches, but direct the growth of the organism. In a certain sense they represent bio-catalysts.

The best known is boron, which regulates part of the physiological processes and, therefore, the state of health in certain plants. The most interesting case is that of heart rot in beet caused by boron deficiency. A small quantity of boric acid (four pounds to the acre) spread on a field prevents this disease. More important probably is the regulation of the evaporation of water from the leaves in summer by reason of the presence of boron. In this case, and especially with clover and grasses, boron increases resistance to drought—a fact of the greatest agricultural importance for the general combating of drought. Once this was discovered it became of interest to determine whether the bio-dynamic products would show similar effects.

Boron influences the respiration process in plant cells. It increases the reduction of the organic substances in the plasma so that finally hydrogen is set free. This can easily be shown in a laboratory experiment since certain dyes like methylene blue are transformed through hydrogen in the status nascendi into colourless substances. Boas has made the following experiment: cubes of equal size are cut from potatoes and put in test-tubes. A few c.c. of methylene blue solution (1 : 50,000) are added. The reduction of the methylene blue through living potato tissue starts and after twenty-four hours the blue colour has disappeared—provided, of course, that only healthy tissue has been used. Even the little cubes which were blue in the beginning have lost that tint. If a 3 per cent solution is added no bleaching occurs. This solution destroys the living tissue without reduction. Boric acid applied in a high solution increases the reduction. The best effect is attained if a 0.0006 per cent solution is used. It is interesting to see that up to this point the more the solution is diluted the greater the bleaching effect.

This simple technique has been applied not only to boron but to the bio-dynamic preparations 502–507. Solutions of from 0.6 to 0.0006 have been used with the result that a strong effect upon the potato tissue was observed. The enzymatic fermentation has been increased with preparations 501, 502, 503, 505, 507, in such a way that it surpassed the best boric acid concentration. Preparations 500, 504, and 506 were second to the best boric acid concentration. A second experiment with 501 and a hundred test-tubes with potato cubes has confirmed this finding. The increase of respiratory effect was 15 to 20 per cent as compared with the best possible result obtained by boric acid. However, a peculiar fact was observed, namely, that the effect of boric acid depends upon the concentration, whereas the bio-dynamic preparations show the same effect with any concentration; that is, the high dilution is just as powerful as the concentrated dilution. This shows clearly that we have here a dynamic effect, because it is independent of the number of molecules present. The bio-dynamic preparations react as true bio-catalysts. Furthermore, the experiment proves that 501 successfully replaces boron.

Another group of experiments dealt with the geotropism and phototropism of plants. Geotropism is the principle that plant roots grow towards the centre of the earth, and phototropism the principle that the shoots grow in the opposite direction, towards the light. Science has discovered that these properties are produced by certain substances present in plants in very high dilutions, namely, auxins and hetero-auxins.

The normal sensitivity of a root—the vegetation point of the root tip contains a very sensitive organ for geotropism—can be paralysed by certain chemicals—eosin being one of them, which in a dilution of 1 : 10,000 so paralyses the orientation of the roots that they no longer grow towards the earth's centre but in any other direction. Only after several hours do the roots recover their original sensitivity.

The auxins are sensitive to light. It is presumed that they are mobilized through light. The paralysing effects are produced by fluorescent dyes which exclude light and therefore reduce sensitivity to light.

According to our experience the bio-dynamic preparation 501 increases all the processes in a plant which have a relationship to light. If there is a disturbance of the geotropism or phototropism, 501 can heal it. The following experiments demonstrate this. Square glass plates, 3 x 3 inches, are covered with a thin layer of moist clay; linseed are introduced into the clay for the purpose of germination. The glass plates stand on cork so that the seeds sprout downward. The whole paraphernalia is put into Petri dishes with moist filter paper to prevent drying out. The experiment has to be made at a constant temperature of 25 degrees. With normal positive geotropism the roots grow vertically downward. When the roots are from 3 to 4 mm. long, the glass plates are turned upside down so that the roots are now forced to grow upward. After about three hours the roots turn like a hook in order to grow back towards the earth—with what is called positive geotropism. If before turning the glass plates the roots are put for only a few seconds into an eosin bath, they completely lose their orientation and grow in any direction. After about twelve hours this paralysing effect disappears and the roots start growing earthwards again. Eighteen hours after the eosin bath they behave normally. The experiment has been divided into two—one as described above, the other so performed that, after the eosin bath, the roots were for fifteen seconds dipped into an aqua solution of 501; 0.5 gramme of 501 has been stirred into three gallons of water such as is always used in bio-dynamic practice.

The result was that the roots treated with eosin and afterwards with 501 did not lose their geotropism. To avoid the objection that the 501 bath washed away the eosin's effect, a bunch of roots was dipped for fifteen seconds into distilled water. The following table gives the final result:

Positive Geotropism

1. Roots untreated	95 per cent
2. Roots treated with eosin	10 per cent
3. Roots treated with eosin and afterwards with 501	87 per cent
4. Roots treated with eosin and afterwards with distilled water	20 per cent

Summary: preparation 501 reacts like light and enables the plant to resist paralysing effects as to geotropism and phototropism.

Since 501 reacts like light, it is permissible to assume that, like light, it stimulates assimilation.[1] Obviously, its effect on plant life being so important, it should be applied at the proper moment. Theoretically, the right moment occurs at the beginning of the growing season just when the lengthening and leaf assimilation start. To find this proper moment experiments with wheat, oats, barley and sunflowers have been made.

Preparation 501 is applied following the usual procedure:
1. With the development of the second leaf.
2. With the beginning of more intensified growth, e.g. with the third stem development.
3. Toward the end of developing new stems.
4. After blossoming.
5. A control without 501 is kept.

[1] Laboratory experiments demonstrating this will be published later on.

SOIL FERTILITY, RENEWAL AND PRESERVATION

For illustration of the effects a few examples are given below:

Experiment with Wheat No. L. 108

Group	Stooling	1000 seed weight in grammes	Yield per plant in grammes	Yield per acre in bushels (approx.)
1.	5	48.4	12.1	36
2.	8	50.0	19.2	56
3.	7	48.0	16.0	47
4.	6	48.0	15.0	44
5.	6	46.0	13.3	38

The best effect is shown if 501 is given at the beginning of the growth of stems. If the application is given too early or not at all the yield is relatively poor. The great difference in the yield indicates the importance of the proper use of 501. Everyone can draw his own conclusions as to whether it "pays" to use 501 or not.

Group	Stooling	1000 seed weight in grammes	Yield per plant in grammes	Yield per acre in bushels (approx.)

Experiment with Oats No. L. 109

Group	Stooling	1000 seed weight	Yield per plant	Yield per acre
1.	3	38.0	16.1	47
2.	4	38.8	19.8	58
3.	3	36.9	14.7	43
5.	3	34.7	18.6	55

Result: same conclusion as for wheat.

The experiments made with 501 showed that this preparation can heal the damaging effect of certain dye substances like eosin. Another bio-dynamic preparation with a similar effect is 507. It is made of Valeriana officinalis, the juice pressed out of the valerian blossom being used. In practice this juice, in a dilution 1 : 10,000, is added to the compost or manure heap in order to transmit its beneficial effect via the humus to all plants. Experiments of the same kind as for 501 have been used for the demonstration of 507 in order to show its healing effect on a disturbed geotropism.

After the plants had been turned completely upside down, the roots restored the disturbed geotropism with the following percentages:

1st group: untreated seeds	94 per cent
2nd group: seed treated with eosin	12 per cent
3rd group: seed treated with eosin, then "healed" with 507	89 per cent
4th group: seed treated with distilled water	23 per cent

The experiment shows clearly the "light" effect of 507.

Bio-dynamic preparation No. 501 is made from quartz (rock crystal) which has undergone a long process of activation. If instead of the treated 501 just the original raw material of quartz is used, only a 34 per cent healing effect is obtained as compared

SCIENTIFIC TESTS

with 87 per cent with the treated quartz (501). The particular bio-dynamic process of activation, therefore, is necessary in order to obtain satisfactory results.

Experiment with Barley No. L. 110

Group	Stooling	1000 seed weight in grammes	Yield per plant in grammes
1.	from 4 to 20	63.5	24.0
2.	,, 10 ,, 22	67.0	36.8
3.	,, 6 ,, 30	60.5	26.0
4.	,, 6 ,, 30	60.5	36.0
5.	,, 4 ,, 22	58.0	24.7

Result: the same as for other grain. In this case the proper application of 501 is remarkably successful. As in other cases the use of 501 during the blossoming period should be avoided.

Experiment with Spinach No. L. 138A

	Yield per plant in grammes, after 7 weeks	Observations
1. Control without 501	14.4	Much blossom started
2. 501 applied when the first leaves developed	14.5	Very little blossom
3. 501 when the third leaf started	15.5	About the same
4. 501 when the plant was well developed	12.4	Much blossom started

Result: the influence of 501 is beneficial as to the formation of green leaves, e.g. for the assimilation process, and delays the plant's going to seed if applied between the formation of the second and third leaves. Going to seed is very undesirable in spinach; 501 keeps the spinach in the more useful state of growth for a longer period.

For the study of the assimilation process sunflowers are very often used. Their large green leaves offer an opportunity for observation and analysis of the growth process of substances like chlorophyll, which are the bearers of assimilation.

The influence of 501 upon the assimilation of sunflower plants has been studied with the following results:

Experiment 1 with Sunflower No. D. 331

	With 501	Without 501
Height of sunflower	8 ft 8 ins.	7 ft. 4 ins.
Number of flowers per plant	18	12
Weight of green leaves per plant	350 grammes	100 grammes

Experiment 2 with Sunflower No. D. 381

Average number of flowers (large) per plant	6.6	2.9
Average number of flowers (small) per plant	3.4	0.7
Seeds already ripe per plant	0.6	4.7
Average weight of green leaves per plant	381.0 grammes	334.0 grammes

Both experiments show an average figure for 25 plants each.

These results are self-explanatory illustrating as they do the superior effect of preparation 501 as regards growth.

Conclusion: 501 should be given at the moment when the plant starts its full development, sufficient root system having been developed. If 501 is given too early, then the plant will develop a strong, juicy, thick root at the expense of the upper parts. This effect, however, could be eliminated through a second spraying with 501 later on. Even if applied at the proper moment, its effect is augmented by a second spraying. Indeed this is especially advisable in wet seasons with little sunshine, and after winters with much moisture and little sunshine. Evidently, in such cases, 501 is a substitute for sunlight. To a certain extent it wipes out the bad effect of too much moisture. In greenhouses where part of the light is absorbed by the glass 501 should be sprayed at regular intervals. If the first dose of 501 is given too late, it forces the plant to mature too early, a thing which does no harm in a wet year but which nevertheless should not be done to lettuce, spinach and other green-leafed plants inasmuch as it causes them to blossom.

Compensation for Lack of Lime with Bio-Dynamic Preparations 500, 503 and 505

The following report deals with growth experiments following the classical scientific method, e.g. in a standard nutritive solution. The growth in such a solution is compared with the growth in another solution consisting of the same ingredients but without calcium. A compensation for calcium has been sought through the application of bio-dynamic preparations 500, 503 and 505. These preparations strengthen the resistance of a plant against calcium deficiency. The following experiment was made in order to demonstrate the influence of a normal and a calcium-deficient solution as compared with the influence of the humus preparations upon germinating plants.

Experiment with maize. Seeds of maize were germinated and, when sprouts had developed 2–3 inches in length, they were placed in the nutritive solution.

The first group contained the so-called Knop Nutritive Solution. In 1000 c.c. of distilled water, 0.25 gramme of magnesium sulphate, 1 gramme calcium nitrate, 0.25 gramme potassium phosphate were dissolved, and 0.25 gramme potassium chloride and a trace of iron chloride were added.

SCIENTIFIC TESTS

Second group: the same nutritive solution, without the calcium, but with the equivalent amount of nitrate nitrogen in the form of ammonium nitrate.

Third group: the nutritive solution as in the second group, deficient in calcium but with the addition of 0.1 gramme to 1000 c.c. each of preparations 503 and 505.

Fourth group: the same as the third, with 1 gramme of preparation 500 to 1000 c.c. of the solution added.

Preparations 500, 503 and 505 contain humus matter, but their calcium content is extremely small. For instance, in preparation 500 it is only 7–1000 of a milligramme, and in preparations 503 and 505 it is only 0.0025 gramme in the amounts used above. After 17 days of growth in the solution, one saw the marked stunting of the maize grown without calcium, while, especially in the fourth group—with preparations 500, 503 and 505—an extraordinary intensification of root formation followed, which indeed far surpassed that of the normal nutritive solution.[1]

Average weight of the plants in grammes after 17 days' growth:

	Sprout	Root
1. Normal nutritive solution	2.9	1.5
2. Nutritive solution with calcium deficiency	1.1	0.6
3. Nutritive solution with calcium deficiency, but with 503 and 505 added	2.6	2.2
4. Nutritive solution with calcium deficiency, but with 503, 505 and 500 added	5.3	3.4

[1] Indications that humus matter, even in great dilution, can compensate for deficiencies, are also to be found in the interesting reports by Niklewski, of the University of Poznan (Poland).

CHAPTER THIRTEEN

Fertilizing; its Effects on Health

Scientific tests have demonstrated a general improvement in the baking quality of bio-dynamically grown wheat in contrast with wheat of the same kind grown under other conditions. This has led to the making of bio-dynamic products, such as special flours and breads; the products raised and made thus bear, in Europe, the trade mark "Demeter".

Through feeding experiments with animals it has been possible to observe qualitative differences. The tests were made with white mice. Three strains of mice were fed with bio-dynamic grain, three with minerally fertilized grain of the same sort. In every generation two pairs were raised to maturity from each strain. Thus in each generation six litters, which had been nourished with minerally fertilized grain, were available for comparison with six litters which had been nourished with bio-dynamically fertilized grain.

The raising of the animals was carried out as follows: at the age of four weeks the young were separated from the mother and segregated according to sex. From this time on the animals were regularly weighed. At the age of nine weeks—when in general they are mature—they were paired, so that with great regularity each new generation was born at twelve-week periods. By this arrangement, the most identical conditions possible were created for these experimental animals.

The food used was Ackermann's Bavarian brown wheat, the seventh sowing of this strain, bio-dynamically fertilized for five years, and the same wheat minerally fertilized with kainit and basic slag in amounts customarily used. All the original wheat used in these experiments came from the same region and was therefore grown under the influence of the same climatic and soil conditions. The feedings were arranged on a sliding scale, based on weight. In addition, the animals received a ration of one

FERTILIZING; ITS EFFECTS ON HEALTH

part milk and two parts water. The milk was boiled for ten minutes.

From the results of the experiments the following points are given. The average number of animals in a litter was:

For those fed with minerally fertilized grain—6.2.

For those fed with bio-dynamically fertilized grain—6.7.

There was thus a slightly larger number of animals born in the case of those bio-dynamically fed.

Average weight of animals	Grain minerally fertilized	Grain bio-dynamically fertilized
At age of four weeks	7.9 grm.	8.5 grm.
At age of nine weeks	17.2 grm.	16.4 grm.
Died at the age of nine weeks or less	16.9 per cent	8.6 per cent

The above figures were an average taken from three generations with a total of 164 animals.

Since the number of deaths in the period from birth to maturity can be regarded as a measure of the powers of resistance, these figures show that the animals fed with bio-dynamically fertilized wheat were notably stronger, although in their weight they were slighty inferior to those nourished with minerally fertilized grain. Regarding the cause of the deaths, it must be added that no sickness arose during the experiments, but that the deaths were due rather to the number of weaklings in the litters which, being retarded in their development, died off.

After this an experiment was undertaken with chickens, which continued for a year, in order to determine the influence of feeding on successive generations.

The experimental animals, white leghorns, were hatched 28th June 1931. On October 19th they were put into two separate, exactly similar pens, B and M, with adjoining runs. There were seventeen hens and one cock in each.

Beginning on that date, pen M was fed daily 100 grammes of minerally fertilized wheat per bird (the Wilhelmina variety). Pen B received the same quantity of bio-dynamically fertilized Wilhelmina wheat grown on identical ground.

Both pens received the same kind of mash and cracked maize (not bio-dynamic) to the amount of 50 grammes per chicken, per day. The only difference in the feeding was in the use of differently fertilized *wheat*.

On February 1st pen M received in addition a daily ration of *ordinary, ground limestone,* used for feeding, from one to two grammes per bird. Pen B received the same amount of *Weleda ground feeding-limestone*.[1]

Pen B began to lay eggs on December 11th, pen M on December 26th.

The differences between groups M and B in the course of the experiment were as follows:

B laid more eggs than M. The individual egg was on the average lighter in pen B than in pen M. But the daily *total weight* was higher for B because of the *greater egg total*.

Average weight of one egg per hen per day:
  M = 59.4 grm.   B = 58.1 grm.
Average total weight of the daily egg production of ten hens:
  M = 427.2 grm.   B = 464.5 grm.

In pen B the chickens stayed outside in the run from one to two hours longer in the evening; that is, their vitality, and with it their liveliness and desire for food, were increased. In rainy weather, the birds in pen M sought the shelter of the chicken house sooner than those of pen B.

Total egg production of ten birds for 9 months:
  M 1495   B 1916
Average per chicken in 9 months:
  M 150   B 192
Weight of the average total daily egg production in grammes:
  M 427.2   B 464.5

A repetition of the experiment in 1933 brought a result similar to the 1932 tests. The egg production of coops of ten chickens for 7 months in 1933 was: for the B coop, 1213 eggs, for the M coop 977 eggs. The average egg weight was 64.4 grammes for B as against 61.7 for M.

In connection with these tests, certain further experiments were carried out which show that the bio-dynamically raised feed was beneficial to the animals in *every respect*. The volume of the egg production was not attained at the cost of the quality of the eggs; on the contrary, the qualitative characteristics corresponded with the improvement in quantity.

---

[1] Weleda ground feeding-limestone is prepared according to a special formula of the veterinary medicine division of the Weleda A. G., Arlesheim, Switzerland.

FERTILIZING; ITS EFFECTS ON HEALTH

Comparative Hatching Test

Forty eggs each of the M and the B chickens were gathered and used for hatching purposes. A "Viktoria" incubator was used for all the eggs, in order to have exactly the same conditions for both groups. The eggs were in it from April 22nd to May 12th. The eggs had identical treatment.

The percentage of M eggs hatched was 35.
The percentage of B eggs hatched was 68.

Test for the Keeping Quality of the Eggs

Quality is not only expressed in taste. Products of poor quality spoil more quickly, because, possessing less active life forces, they decay more easily. Hence, in order to test the quality of the M and B eggs, the following experiment was carried out: beginning on April 10th, ninety eggs each from the M and the B coops were gathered and stored up in a dry room in sawdust, under identical conditions. After 2, 4 and 6 months respectively, thirty eggs from each group were tested for their usability. It was evident in this testing that there was a very notable difference between the two groups, as shown in the following table.

The numbers of eggs spoiled in each lot of 30 eggs, taken from each group respectively, were:

	M		B	
After 2 months	4 eggs = 13 per cent		1 egg = 3 per cent	
After 4 months	14 eggs = 47	,,	6 eggs = 20	,,
After 6 months	18 eggs = 60	,,	8 eggs = 27	,,

Effects on Plant Growth

Every farmer or gardener knows how important animal manure is as a fertilizer. He knows, too, that the dung of different animals, cow, sheep, chicken, etc., has different effects on plants. This difference affects the size of the plant and the rate of growth, as well as the taste, quality of the fruit or vegetable, and other characteristics.

There is, however, not only a difference between the manures of various animals; the manures of animals of the same sort show different effects in accordance with the origin and quality of their food. The manure is a mixture of the decomposed products of elimination and the digestive juices, the latter being rich in hormones and ferments. In this there still remains, to an observable extent, something of the qualities contained in the foodstuffs.

SOIL FERTILITY, RENEWAL AND PRESERVATION

Hence, the effect of a manure on plant growth depends—apart from the question of the health of the animal as such—also on the fodder of the animal. "Who feeds well manures well", runs a familiar old peasant adage.

There is even an observable difference in the effect of the manure of animals which have been given the same kind of *feed*, which, however, has been *grown in various ways*. A difference is also to be seen when the manure used is given special treatment, for example, when it is bio-dynamically prepared. The following test shows a difference resulting from different feeding regimes.

Growth Test with Dwarf Beans
Average values for 12 plants in each case, given in terms of one plant:

	M	B	Control
Length of sprout in cm.	49	49	42
Weight of plants exclusive of beans in grm.	58	77	44
Weight of beans: pods more than 10 cm.	79	83	29
Weight of beans: pods less than 10 cm.	4	9	6
Number of beans: pods more than 10 cm.	14	15	7
Number of beans: pods less than 10 cm.	7	11	11
Length of pods in cm.	14.7	14.5	12.2

Taste of Beans
(Tested by 8 persons; judgments were unanimous)

	Raw		Cooked
M	very watery	M	good
B	firmer, more character	B	best
C	dry, insipid	C	poor

From the chickens of the M and B groups, the manure was gathered and stored up separately and then mixed with earth of the same sort and set up in separate compost heaps. This was done on 2nd February 1933. Both compost heaps lay side by side under the same conditions of light and warmth and had the same bio-dynamic preparations inserted in them at the same time, March 14th.

Garden beds were fertilized with these chicken manure composts on July 24th. One bed was fertilized with M compost, the other with B compost, a third was left unfertilized as the control C. The plant seed, sown in the three neighbouring beds, was of exactly the same sort, and was allowed to grow under ordinary conditions. Radishes and the dwarf beans already mentioned were used as experimental plants.

The preceding and following tables concern these experiments; they show in every case the same results and differences, not only in quantity but also in quality, as was shown in the flavour test.

FERTILIZING; ITS EFFECTS ON HEALTH

Growth Test with Radishes
Average values for 16 plants in each case, given in terms of one plant:

	M	B	Control
Leaf length in cm.	12.6	9.5	9.6
Leaf weight in grm.	4.1	5.0	4.0
Number of leaves	4.5	5.6	5.6
Length of root in cm.	2.3	2.6	2.2
Weight of root in grm.	7	8.4	6.3

Taste: B   strong, not sharp
        M   medium
        C   mild, too dry

A like qualitative difference between the two chicken manure compost heaps was also evident when the same experiment was made by the "method of sensitive crystallizations".[1]

A very important relationship was discovered here. An original quality of the fodder—its nourishment value—persists and influences the capacity for work of an animal; but in turn the quality of the manure itself is influenced and produces again new growth and new enhanced qualities as the seed, feed and fertilizer cycle continues. Thus, the bio-dynamic cycle leads to a constant improvement in all the elements concerned until the highest degree of performance and health is reached in every organism involved.

It is, furthermore, interesting to note that animals react exactly to variations in the quality of their food. Grazing cattle select certain plants and herbs and avoid others. The peasant, who knows his animals, knows also what they like to eat. He sees how the cow avoids the plants of the crowfoot family (genus Ranunculus) which are poisonous in their green state. He is afraid of the white hellebore and meadow saffron because these do not lose their poisonous quality even after being cut in the hay. The meadow foxtail, cocksfoot and meadow fescue are well liked by cattle, while the broom sedge and common sorrel are eaten unwillingly or not at all. With sure instinct the animal on the pasture chooses those plants which are most beneficial to it, and where the farmer is sowing pasture he will try, on the basis of such observations, either made by himself or coming from others, to bring together the plants that are gladly eaten by his animals. Due to economic conditions and also to the status of science to-day (it is only necessary to think of the widespread, year-round stabling of animals),

---

[1] E. Pfeiffer, *Formative Forces in Crystallization*, London, Rudolf Steiner Publishing Co.

SOIL FERTILITY, RENEWAL AND PRESERVATION

it is not always possible for the individual farmer to cultivate these intimate observations and their place has to-day often been taken by scientific experimentation.

The following experiments may help to clarify this question. White mice were used. These are in general preferred as test animals because of their sensitivity. In the mouse, the sense of taste and of smell are extraordinarily highly developed. It is an animal—extremely nervous and sensitive—which possesses organs that react strongly to changes in its nourishment.

The experiments were carried out in such a way that the animals had two similar dishes of wheat set before them. This wheat was of the same variety, and had been grown for five years on the same soil just for experimental purposes, but the wheat in one dish had been treated with mineral fertilizer (calcium nitrate and ammonia plus "Leuna"—a trade name—potassium nitrate), while the wheat in the other had been treated with bio-dynamic manure. The wheat in both cases was used in quantities that permitted the presence of the same percentage of nutritive substances. Basis: 12 tons of manure to the acre.

Experiment "A" was made with Bavarian brown wheat, and experiments "B" to "E" with the variety Karsten No. 5. By weighing daily the uneaten feed, it was determined for that day how much the animals had eaten of the minerally fertilized, and how much of the bio-dynamically fertilized grain. In order to obtain unquestionable results, it was necessary to pay attention to a number of things that might appear of secondary importance. The two dishes of food were set up in exactly the same amount of light. The dishes were often interchanged in their positions in order to avoid the objection that the animals would always take either the left or the right dish simply out of habit. In order to avoid the spilling or loss of any grain, neither was the cage in which the animals were housed too small nor was the number of animals used in an experiment too great. By careful inspection and care of the cages it was possible to keep all mistakes that might creep in in that way down to a minimum.

The grain used was harvested under exactly the same conditions, so that it was at exactly the same state of dryness. As supplementary food we used boiled milk, mixed with water in a 1 to 3 ratio for all the animals. If these points are observed, exact results may be looked for.

The experiments were made either with animals which previously, for six or seven generations, had been fed entirely on

FERTILIZING; ITS EFFECTS ON HEALTH

bio-dynamically fertilized wheat, or with other animals which had been fed only minerally fertilized wheat. Table "A" shows such a food-selection test with female animals:

Table A	Bio-dynamic Wheat in grammes	Minerally fertilized Wheat in grammes
Mice bio-dynamically pre-fed for 6 generations eat in 24 days	108	10
Mice minerally pre-fed for generations eat in 24 days	98	5

In this experiment it soon becomes apparent that bio-dynamic feed is eaten almost exclusively. The animals tried the mineral wheat only at the beginning and then left it untouched in the trough. It might be objected that the animals were accustomed to a certain kind of food through inherited habit, if a still more one-sided result had not shown itself in a family of mice which had been altogether "minerally" nourished through six generations. These animals dropped their habit of eating mineralized food completely, and instead immediately chose bio-dynamically fertilized grain whenever they were given the opportunity. The other kind of grain was only occasionally eaten.

That an excess of salts, in the case of mineral fertilizing, can have a harmful effect on the food, has been shown from various studies. Reference has already been made to the observations of the Swiss research worker, Dr. v. Grünigen (page 56). They point to the danger of a too high potassium content; this comes from the tendency of the plant to consume an overabundance of potassium. The data of Professor Rost of Mannheim[1] go further in presenting the danger of a surplus of potassium. Rost demonstrated by feeding experiments that, through the feeding of potassium, thrombosis as well as gangrenes could be experimentally produced. To quote from his important treatise: ". . . But now I made, in connection with the potassium-nitrate fed animals, an observation which is extraordinarily interesting, for they showed a tendency, a pronounced inclination in the successive generations, towards thrombosis. . . ." Obviously these phenomena appeared more pronounced in the second generation than in the maternal animals. Rost testifies that, in recent years, thrombosis has in-

[1] Professor F. Rost, "Ueber Schwanz-und-Fussgangraen bei Ratten", *Munchener Medizinische Wochenschrift*, 1929, No. 22.

SOIL FERTILITY, RENEWAL AND PRESERVATION

creased also in the human being up to four times its earlier prevalence, and he comes to the following conclusions:

... The potassium content of plants may be considerably increased by potassium fertilizing. ... In cooking, the chief part of it goes over into the water. Using the figures given in König's *Chemie der Nahrungs- und Genussmittel*,[1] one can reckon that, for example, with spinach, about two-thirds of the mineral substances goes over into the cooking water and is poured out with it.[2] But lately there is a strong tendency not to pour off this cooking water as was formerly done, but to utilize it, because these mineral substances are considered something especially important for human nourishment. Without doubt we absorb considerably more potassium salts with this modern way of cooking than formerly, to which must be added the fact that, thanks to the plentiful use of artificial fertilizing, the potassium content of the plant is higher than in earlier decades. I may, perhaps, interject here that this modern way of cooking by no means agrees with everyone's digestion. I know many otherwise completely healthy people who are affected with nausea and severe diarrhoea as a result of partaking of vegetables on which the cooking water has been retained, and it is further known that the "pollakiuria", which was so often observed during the World War, was mostly a "polyuria" resulting from potassium salts.[3]

It is very natural for us now to underline these sentences and to conclude that the increased tendency to thrombosis, as we have observed it in recent years, stands in direct relationship of cause and effect with the increased potassium content in food. I personally am of the opinion that we should be very careful about drawing such a conclusion in so definite a form. ... We may indeed always speak of the fact, that according to the animal experiments referred to, and the other data given concerning potassium feeding, such a conclusion is completely tenable and justified when considering the whole thrombosis question from this viewpoint. Until now we have been unable to give any well-founded explanation for the increase of thrombosis in recent years. Hence, we may certainly regard this as an encouraging result of the animal experiments cited, and be confident that they indicate to us the direction to follow in our researches into this highly important question.

Furthermore the experiments are of interest because it was possible, through doses of salts which apparently were quite harmless to the animal itself, to bring out states of sickness in the second generation.

---

[1] König's *Chemie der Nahrungs- und Genussmittel*, IV Aufl., Bd. 2, p. 1458.
[2] There was contained in:
   1. 100 grm. spinach, raw and unwashed, 0.695 grm. potassium.
   2. 100 grm. spinach, raw but washed, 0.0602 grm. potassium.
   3. 100 grm. spinach, cooked, cooking water poured off once, 0.192 grm. potassium.
   4. The water poured off from 3. contained 0.379 grm. potassium.
   5. 100 grm. spinach, cooking water not poured off, 0.601 grm. potassium. Thus cooked spinach from which one does not pour off the cooking water contains about *three times* as much potassium as spinach prepared by the old method.
[3] Unpublished research.

FERTILIZING; ITS EFFECTS ON HEALTH

The experiments of Tallarico,[1] too, prove the influence of fertilizing on the quality and the health-giving properties of foods. Note the following:

> I was able to make certain observations in a series of experiments with grain —which had to do with the fertilizing and development of the mother plant, as well as with the capacity for yield of seeds thus produced. There was a noticeably different behaviour in seed coming from minerally fertilized mother plants compared with seed coming from mother plants which had been fertilized with stable manure. While the first group in general gave a modest yield, the second, which was cultivated under identical conditions of soil, climate and agricultural procedures, gave a more luxuriant growth and a higher yield.

Turkeys are very well suited for experimentation of this sort for they eat practically anything and mature quickly. They are above all useful in this way because with them the time of puberty brings the so-called "red crisis", whereupon the characteristic red growths appear on the head and neck of the bird. During this critical period of development the animal gets into a state of profound weakness, in which it easily succumbs to intestinal or lung infections, which bring trouble even to the best flocks. This natural sickness, which only exceptionally spares the strongest, comes more or less early, lasts for a longer or shorter time and ends in death or recovery, in accordance with the greater or lesser powers of resistance of the animal attacked. In order to estimate these powers of resistance an experiment was made in which the number of ill birds in each test group was observed and the time of the beginning of the crisis, its length in terms of days, and the manner of its termination were noted. In this experimental flock there were also cases of birds which, as a result of a real and genuine recovery from the disease, remained small all their life; in other cases they remained so greatly weakened that they had to be taken out of the flock, because they were unable to obtain their share of the food. These birds were given the special category of "stunted". The following is the report of the experiment:

"The food was raised on three parcels of land. The *first parcel* was fertilized with mineral fertilizer each year for two years, with 180 lbs. of ammonium sulpho-nitrate per acre, spread during the time of cultivating the field, and also 90 lbs. of potassium per acre

---

[1] G. Tallarico: "The Biological Value of the Products of Soil Fertilized with Animal or with Chemical Fertilizer". *Proceedings of the R. Accademia Nazionale dei Lincei. Mathematical, natural-scientific division.* Vol. XIII, series 6, 1. Rome, February 1931.

and 350 lbs. of super-phosphate per acre, which were spread after seeding.

"The *second parcel* was fertilized each year for two years with rotted stable manure, in a proportion of 1,000 lbs. per acre.

"The *third parcel* was used in its natural state, without mineral or animal fertilizer.

"On these three parcels, the most important foods for the raising of the turkeys were grown in separate beds. In the second year, a portion of the produce grown on the same parcel the first year was used as seed.

"The foods, thus grown, were supplemented with egg yolk and ground meat, and formed the four feed groups which were given to the four groups of experimental animals in the first six months of their lives, in the form of a mash, which was composed of equal parts of the various ingredients by weight:

"Type A: meat residue, plus egg yolk, plus grits from stable-manured grain,[1] plus whole, stable-manured grain, plus stinging nettle and cut leaves of sweet clover, both the latter from parcels which had neither synthetic fertilizer nor stable manure.

"Type B: meat residue, plus egg yolk, plus grits from minerally fertilized grain, plus minerally fertilized whole grain, plus stinging nettle and sweet clover leaves from unfertilized plants (feeding with minerally fertilized grain).

"Type C: meat residue, plus egg yolk, plus grits from unfertilized grain, plus unfertilized whole grain, plus stinging nettle and cut leaves of sweet clover, both from parcels which had been fertilized with stable manure (feeding with stable-manured green feed).

"Type D: meat residue, plus egg yolk, plus grits from unfertilized grain, plus unfertilized whole grain, plus stinging nettle and cut sweet clover leaves from parcels which had been fertilized with mineral fertilizer (feeding with minerally fertilized green feed).

"In order to determine the organic capacity for resisting disease, the following phenomena were observed and considered: the number of sick animals, the beginning of the crisis for each animal, its length and its termination, whether in death, cure, or stunting. From these results the mean percentage was then determined for each group and each test series, taking into consideration

---

[1] The minerally fertilized, the stable-manured and the natural grain were all given to the animals in the form of crushed grain in the first month and as whole grain in the second month, the time when young turkeys begin to take whole grain.

FERTILIZING; ITS EFFECTS ON HEALTH

the unavoidable accidental losses in raising them—such as death through cold, injuries, or birds of prey.

"From this it is plain that the turkeys which were fed in the first two months with grain or green feed from plants grown on stable-manured soil show, when they are attacked by the crisis, a smaller number of cases of sickness, a shorter duration of it, and a lesser number of cases ending fatally, in comparison with the corresponding group of turkeys which under the same conditions of life and environment were fed with green feed or grain from plants that were fertilized with mineral fertilizer.

"In one table are gathered together the average mean percentages obtained in each group for the three experiments:

"Organic powers of resistance in young turkeys

Feeding with:	Number attaining crisis per cent	Age at start of crisis days	Duration of sickness days	Result: died per cent	cured per cent	stunted per cent
A: stable-manured grain	89	43	7	21	79	0
B: minerally fertilized grain	94	47	11	34	64	2
C: stable-manured green feed	82	39	6	18	82	0
D: minerally fertilized green feed	96	46	10	39	60	1

"Furthermore, it is to be gathered from this, that the leafy growth of plants fertilized with stable manure has a more beneficial influence on the various phenomena considered here than do the reproductive organs of plants which are similarly only fertilized with stable manure.

"*This means that the seeds, and still more the leaves, of plants fertilized with stable manure have the peculiarity, when used as food for these animals, of increasing their capacity for resisting disease to a greater degree than the corresponding seeds and leaves of minerally fertilized plants. The former have thus a higher biological value than the latter.* This conclusion is also confirmed by the lack of cases of arrested development and stunting in the groups which were nourished with products fertilized with stable manure."

This presents an extraordinarily important aspect of the bio-dynamic method of agriculture. It is not only able to improve the soil in its organic structure; its consequences reach far into the kingdom of man. If in only one instance the influence of various methods of agriculture on the health of animals and man as it is presented here is observed, then it is to the interest of every *consumer* to concern himself with the sources of the food he eats. *He can and should demand of the farmer that he furnish him the maximum of health-giving qualities in his bread, vegetables and fruit. The consequences for hygiene and health of such a stand on the part of the consumer are incalculable.*

Experience in this respect has always shown that, wherever the bio-dynamic method of agriculture has been used, the attention of physicians is soon directed to the effects of the products. Hence, we shall cite several reports in this connection. The reader may see from them that agriculture is not only a concern of the one who cultivates the soil, but that every human being may well be concerned with respect to the manner in which he nourishes himself.

J. Schulz, M.D.,[1] testified that it was possible for him, with the help of bio-dynamically raised food—in the form of bread—to cure a series of metabolic disturbances, and on the basis of this diet to obtain a stronger effect with his medicinal therapy. He observed these beneficial results in children as well as in adults.

R. Reinhardt, M.D.,[2] and J. Kalkhof, M.D.,[3] gave similar observations. Another physician states[3] ". . . that bio-dynamic products are necessary for therapeutic control of diets and cannot be replaced by other products on the food market. As a physician I acknowledge gladly that for weak and backward children especially such food is necessary. . . .

"Gradually we have gone over to using the bio-dynamic products, which appear to be of good quality and have a definite influence on the functioning of the stomach and intestines. I have recommended these products to patients with marked stomach troubles and sluggish intestinal activity, and they have been fortunate enough to get over these ailments without medical treatment." Further ". . . my wide experience as a dietician with many patients has convinced me that especially with a raw food diet the bio-dynamically treated products are preferable in every way to those which have been manured in the usual way with chemical fertilizer or by the use of faeces."

[1] "Diäterfahrungen mit Demeterprodukten der biologisch-dynamischen Wirtschaftsweise" in *Fortschritte der Medizin*, 7th January 1935, Berlin.

[2] R. Reinhardt, M.D.: "Einiges über Ernährung unter Berücksichtigung des Zusammenhanges von Ackerboden, Pflanze, Tier und Mensch". *Hippokrates*, Zeitschrift für praktische Heilkunde, Vol. 5, No. 10, Stuttgart.

[3] J. Kalkhof, M.D.: "Beobachtungen und Krankenerfahrungen mit Demeter-Ernährung". *Aerztliche Rundschau*, No. 21, 1935.

FERTILIZING; ITS EFFECTS ON HEALTH

Physicians' Reports on their Use of the Bio-Dynamic Products. Cases a, b, c, d

(a) "The excellent results which I have had in the use of bio-dynamic products—used in connection with the treatment of an ever-increasing number of my patients—have convinced me of the urgent need at the present time of having an unlimited supply of these products which are of medical hygienic value for the general food supply."

(b) "I am glad to state my appreciation of the value of the 'Demeter' products. I view them as a *model* to be followed at a time when the inner value of many of our foodstuffs is declining. It seems to be necessary in the interest of public health to have these products first on the markets, then later to introduce them into general use for everyone."

(c) "I have no opinion concerning the cultural background of these products—but I endeavour to promote their use because of their beneficial effect on digestion. I try to advise my patients suffering from certain forms of digestive disturbances to use them. By using them I have in many cases been able to make cures without drugs."

(d) "I do not hesitate to state that in my opinion the bio-dynamic products are of the greatest value for public health—especially in diseases requiring dieting. My experience as a dietician with numerous patients convinces me that especially raw vegetables and fruit grown in the bio-dynamic method are more healthful than are the products of other methods of agriculture."

In every case it was observed that, with a change-over of diet to bio-dynamic products, there was at first an immediate increase of appetite, which during the first weeks resulted in an increase in the consumption of food. After two to three weeks, however, a state of equilibrium occurs, with the final result that the person has enough to eat with only two-thirds of the volume previously believed to be necessary. This fact is also a proof of the increased nutritive value of these foods. Observations in this field have been made extensively with hundreds of test cases, over a period of years.

The well-known physiologist, Geheimrat Abderhalden, takes the following position with regard to the problem: "In connection with various illnesses of man and animal it has frequently been desirable to trace them back to the method used in fertilizing the food plants. Nothing can yet be said with any certainty, but we must keep in mind the fact that important substances come from soil bacteria, and we must consider whether it is correct to disturb the fine interplay of all the soil organisms by bringing in nitrogen in the form of potassium nitrate and by using lime and phosphoric acid, because the development of the various sorts of organisms is thus disturbed and hindered, and on this account difficulties will some day arise." In another place Abderhalden says: "If we take care of the soil exclusively with chemical fertilizer, then it is indeed conceivable that disturbances in the growth of the plants will occur. It is especially conceivable that the development of the unknown substances (vitamins) may be menaced. That organism, 'the soil', will in a definite way find itself in the same condition as an animal which receives the materials of nourishment only in their chemically purest form. Obviously it becomes ill. This organism, 'the soil', with its concourse of cells and their manifold reciprocal effects, will quite definitely become sick." (Cf. the previously cited article of Dr. Reinhardt.)

Let us, before leaving this subject, cite a British research report.[1] R. McCarrison, B. Virwa Nath, and M. Suryanarayana are joint authors of a noteworthy study on the influence of chemical and organic fertilizing. They found very important qualitative differences in the case of seeds of the millets, **Eleusine coracana** and **Panicum miliaceum**, and of wheat. The differences were traced as far as feeding tests. Under the influence of the warm climate these grains give greater yields with organic fertilizer than with chemical fertilizer or without any fertilizing. In contrast to the yield without fertilizing, chemical fertilizing brought a yield increase of 32.8 per cent and organic fertilizing an increase of 100.7 per cent in the case of Panicum miliaceum. The same strain of seed was used again and again for the same fertilizer test. Thus an increase was attained in the various qualitative effects. Feeding tests with Eleusine, using pigeons, gave the following results:

	Average percentage of loss of body weight during the days of the test
Group with basic ration	37.7
Group with basic ration plus plants grown on stable-manure basis	22.4
Group with basic ration plus chemically fertilized plants	37.4
Group with basic ration plus unfertilized plants	40.9

[1] *Memoirs of the Department of Agriculture in India*, IX, No. 4, 1927, B. Virwa Nath, M. Suryanarayana, R. McCarrison. Cf. also *Demeter*, 1934, No. 12, F. Dreidax, "Das Mark der Landwirtschaft".

FERTILIZING; ITS EFFECTS ON HEALTH

"Even with manifold changes in the conditions of the experiment, a better result was evident in the case of the seed raised with organic fertilizers than with seed raised with chemical fertilizers. In the case of the wheat, the seed raised with chemical fertilizers reacted less favourably than the seed raised without fertilizer."

Feeding Tests with Pigeons and Barley
(in terms of loss in body weight)

	per cent
Stable-manured barley showed better results than the "unfertilized" by	18.5
Stable-manured barley showed better results than the "minerally fertilized" by	15.0

Experiments with rats, with a basic ration of meat residue, refined starch flour, olive oil and salt, and cod-liver oil (or Marmite or Vegex) for a vitamin supplement, and, besides the foregoing, either organically fertilized wheat or minerally fertilized wheat, both sorts having been grown on adjoining parcels, gave the following results:

	Percentages of gain in bodily weight:
Basic ration plus stable-manured wheat	114
Basic ration plus chemically fertilized wheat plus vitamin supplement	104
Basic ration plus chemically fertilized wheat alone	89

That is, the stable-manured wheat is even better despite the sharp competition of the vitamin supplement. We are surprised that these important experiments have so far been little quoted in scientific literature. It is to be assumed that such experiments run counter to so many fondly held dogmas in the spheres of agriculture and nutrition; but that ought not to prevent repeated objective examination of this question.

When we also consider in this connection the researches of Professor Boas of Munich,[1] who proved that pasture grasses organically fertilized possess a higher albumen content than such grasses minerally fertilized, which, however, contain more peptone (the first product of the disintegration of albumen), we are justified in the opinion that, when these facts are more widely known, people will manifest a greater interest in the problem of quality in agricultural products than is now the case.

[1] Professor F. Boas, "Untersuchungen für eine dynamische Grünland Biologie", *Praktische Blätter für Pflanzenbau*, IX, p. 173. 1932.

CHAPTER FOURTEEN

Practical Results of the Bio-Dynamic Method

Theory and principle may be never so sound and correct, yet it must always be asked: do they work out well in practice? In this scientific age there is still another question: do they prove themselves in scientific experimentation? The scientific experiment has one peculiarity: it aims to be as exact as possible. Hence all experimental errors must be eliminated as far as this is possible. This necessitates the narrowing down of the experiment to as small a compass as possible, while still retaining the functioning power of the principles to be tested.

If such a complicated process as plant growth in its relation to soil and climatic conditions is being dealt with, it is difficult and almost impossible to keep an accurate check on *all* the factors. Thus one is usually satisfied, in such experiments, to set up and test a series of detailed minutiae. Then the whole is assembled from the parts. The question is whether one then still has contact with reality—that is, is one dealing with practical things?

In tests of the bio-dynamic method, we have also done this same sort of thing. Elsewhere, in reports of scientific experiments, we have presented the results of one or another of these experiments by means of which a small segment of a whole problem could be shown. Yet it is quite generally recognized to-day that, metaphorically speaking, the most exact knowledge of what takes place, for example, inside an internal combustion motor is by no means enough to enable anyone to build a really good petrol engine.

In an attempt to get closer to agricultural practice, we have directed the scientific method of experimentation towards something which apparently is within the sphere of real practice—namely, the soil. Thus we made "comparative experiments" in this field. Parcels of land lying side by side were repeatedly cultivated in the same way, planted with the same crops, except that they were fertilized differently to demonstrate the differences in

PRACTICAL RESULTS OF THE BIO-DYNAMIC METHOD

fertilizers. Such experiments were then carried on for one, two, three, and sometimes even for four years. The author has himself, in the past, started and arranged such experiments, but was forced to realize that they could give only an inconclusive picture of the biological alteration of the soil. Yet this biological transformation is the very foundation of the practical value of the new method of agriculture. Such experiments can be conclusive only when the soil has not been rendered incapable of biological activity as the result of previous abnormal treatment.

The biological enlivening of the soil, the changing of its structure, and even certain qualitative improvements in it all require time. The more any one of its factors has strayed from its natural basis, the longer is the time required to get back to normal healthy conditions. But this period does not belong to the experiment proper, but *only to the preparation for it.*

In an agricultural experiment it is only when we get back, in the course of the crop rotation, to the same fertilizing and the same crop on the same land in which it was planted before that the "experiment" really begins. Then the soil has had time to develop, and the seed from the first crop can be used for the second. Then we know the peculiarities and special requirements of that particular soil, so that we can cultivate it in such a way that the life stimulated in it can really come to expression. The author once had occasion to carry on a comparative experiment which was discontinued after three years as having been negative in its results. In this experiment certain essential points had been ignored: (a) the influence of a crop rotation, (b) the fact that the experiment really should not have been begun until the third year after the field had been converted, (c) also the special fact, that the land had been used for decades as a military exercise field. Only within the last ten years had it been turned into farmland. Because this soil had been an exercise field for decades, it certainly needed a number of years before it could be returned to a "normal and natural" state. This means that we must take into account the time required for a proper preparation of the ground.

In another case the experimental plot had previously been intensively used for a number of years for mineral fertilizer experiments. Here again there had to be a wait of several years before a normal state could be established.

We have frequently observed that the bio-dynamic seed used in making experiments has shown itself to be more resistant to plant diseases; and also that there is an increase in the effect under observation when there are a number of consecutive repetitions of the same experiment in the same place.

Apart from all this, it is evident that the bio-dynamic method of agriculture does not merely result in fertilizing the soil. When properly practised it gives due weight to every factor required for healthy plant growth. We see from this why fragmentary tests, made by using comparative parallel strips of land, are one-sided and hence can lead to no useful results—that is, they must end indecisively.

The ideal experimental basis is, indeed, the *practical farm* on which the results are checked and controlled over a period of years. Such a basis, while it does include all the sources of experimental error found when dealing with nature rather than with an artificial grouping of factors is at the same time the only correct one because it takes into account the full and normal effects of nature, as opposed to an unbalanced, compressed nature.

Complicated as this may appear at first sight, it is only in this way that everything can be done thoroughly and compared exactly. Here we can produce conditions which really have significance for practical application. If an exact crop rotation is determined for certain fields in which the dung of the bio-dynamically fed animals is brought to the bio-dynamic parcel of land, or the dung from the minerally fed animals is brought to the mineral-fertilized parcel, and if the products of these parcels are then given to the animals belonging respectively to these parcels, so that gradually a closed cycle of substances is obtained, then out of practical experience one can judge, from all its effects, whether a certain method of agriculture has value or not. Curiously, however, it is the scientists themselves who have a great hesitancy in attacking this problem which is indeed the decisive one for the practical farmer or gardener, namely: *"What is a method's value when used on a farm over a period of years?"*

To answer this question, we present here extracts of reports from a number of exactly observed and controlled farms, with figures on yields before and after their conversion to the bio-dynamic method. Farm A.: 450 acres of heavy and medium clay soil; chief crops, grain and sugar beet; had had intensive mineral fertilizing before conversion. The converting was done during the years 1922–4 without the experience since accumulated for carrying it out.

The following table shows the usual crop rotation *before the conversion*. Manure was always given to the beet and potatoes.

PRACTICAL RESULTS OF THE BIO-DYNAMIC METHOD

	Four Year		Three Year		Three Year
a.	1st yr. beet	b.	1st yr. beet	c.	1st yr. potatoes
	2nd yr. wheat		2nd yr. wheat		2nd yr. oats
	3rd yr. beet		3rd yr. rye		3rd yr. rye
	4th yr. wheat		1st yr. beet		1st yr. potatoes
			2nd yr. wheat		2nd yr. oats
			3rd yr. rye		3rd yr. rye

Crop succession after the conversion

Four Year

a. 1st yr. sugar beet plus bio-dynamic stable manure
    2nd yr. field peas or broad beans plus oats
    3rd yr. wheat
    4th yr. rye or winter barley, with legumes (hop clover) immediately following

Five Year

β. 1st yr. potatoes plus bio-dynamic stable manure
    2nd yr. wheat
    3rd yr. oats plus broad beans
    4th yr. clover
    5th yr. rye followed by legumes
    1st yr. (new rotation) potatoes plus bio-dynamic stable manure

Average grain yield, in lbs. per acre:

1914—1746	1926—1675
1915—1817	1927—1764
1916—1834	1928—2117
1917—1755	1929—2205
1918—1975	1930—2117
1919—2006	1931—2090
....	1932—2416
1923—1587 (conversion begun)	1933—2160
1924—1323	1934—1605 (long, heavy drought)
1925—1675	1935—2090

Average pea harvest before the conversion, in lbs. per acre (peas before conversion were an uncertain crop): 529–882
Average pea harvest after the conversion: 1764–2028
Average bean harvest after the conversion: 2998–3351

One of the objections frequently voiced by opponents of the bio-dynamic method of agriculture is that things may, indeed,

SOIL FERTILITY, RENEWAL AND PRESERVATION

go well for a number of years, but after a longer period of time the soil will, nevertheless, be exhausted. In 1932, ten years after the conversion was begun, soil samples were sent to a scientific institution. Its report stated:

"The soil is, at least at this time, *sufficiently* provided with potassium as well as with phosphoric acid."

1934: "*Abundantly* provided with potassium, *abundantly* provided with phosphoric acid."

1935: "The soil is *abundantly* provided with potassium, *well* provided with phosphoric acid."

Sugar-beet yield, in lbs. per acre:

1923—18230		1930—26638	
1924—22154		1931—26898	
1925—27710		1932—27714	
1926—22308		1933—28248	
1927—23693		1934—31740	
1928—24914		1935—25029	(lasting drought)
1929—25025			

There was constantly observable a slight increase in the sugar content as compared with the normal mean yield. This result has been repeated on various farms.

	per cent		per cent
1927 sugar content of the bio-dynamic beet	17.2	Mean figures for the sugar-beet factory	15.27
1934 sugar content of the bio-dynamic beet	18.2	Mean figures for the sugar-beet factory	18.14

Potato yield, in lbs. per acre:

	Before		After
1917	10423	1931	20370
1918	8907	1932	17637
1919	7337	1933	17637

The consumers especially praised the keeping quality and flavour of the bio-dynamic potatoes.

Milk production, in lbs. per year per animal:

1914–15	5924	1924–25	6581
1915–16	6433	1925–26	6938
1918–19	4877	1926–27	8245
1919–20	6541	1927–28	8208

1928–29	7412	1932–33	7114
1929–30	7114	1933–34	6219[1]
1930–31	7421	1934–35	6429
1931–32	7811	1935–36	6479

There was an average of 30–35 cows kept on the place; they were chiefly fed with fodder grown on the farm itself. Up to 1928 there was brought in from outside 7 lbs. of peanut meal and sprouted malt per animal per day; up to 1934, 3½ lbs. per animal per day; in 1937 1¼ lbs. per animal per day. The farm had one head of cattle for every 5 acres of arable used land.[2]

Especial emphasis is placed on feeding the cattle with home-grown fodder. We have seen that a mixed farm (with cattle and tilled fields) has greater lasting powers of production and of resistance to harmful effects than any other. The forced feeding of cows to attain a record production of milk has fortunately gone out of fashion again. A natural cattle husbandry, with meadows and hay and clover, will always form the healthy foundation of a farm. It furnishes the manure that is necessary for field culture while the field culture in turn gives over a portion of its produce to the animal husbandry. The "closed circle of feeding" (home-grown feed fertilized with home-produced manure) is the basis for a healthy stable. The fewer cattle replacements, the more home bred, the better.

We encounter however to-day breeding diseases, contagious abortion, mastitis, etc. These epidemic diseases are so familiar that we need only refer here to their existence. Yet we must listen to testimony out of the field of practice.[3] *Farm B.* "At the time we took over the farm, contagious abortion had been present there over a period of years. Through the planned conversion in feeding and the animal husbandry in general, and with the help of the Weleda remedies,[4] we were able to eradicate the contagious abortion from the stable. In the past year, our only loss was one calf, from diarrhoea. All the cows calved 'normally'."

Farm C. "The herd had been suffering from miscarriages for

---

[1] Abortion (Bang's disease) brought in from outside, and slowly eradicated from the herd. 1936 shows further progress after the conclusion of this table.

[2] "Die Entwicklung des Klostergutes Marienstein bei Nörten (Hannover)". *Demeter,* 1936, No. 7.

[3] Dr. N. Rehmer: "Erhöhte Milch-und-Fettleistung durch individuelle Betriebsgestaltung". *Demeter,* 1937, No. 1.

[4] *Weleda, Inc.,* New York City, U.S.A. The *Weleda Corporation,* London, England.

fifteen years, and had been practically annihilated. At this point, the bio-dynamic method of agriculture was introduced; the feeding plan was basically rearranged; the pasture facilities were enlarged; the fodder was practically all raised on the place; and the feed-growing areas were put under intensive bio-dynamic cultivation. The new feeding basis and the treatment with the Weleda remedies made possible an early, complete cure of the herd."

Farm D. "Regularly in the autumn, at the time of winter stabling, and in the spring, there were present both calf paralysis and pneumonia. The losses ran as high as 30 per cent. For the past two years the farm has been worked bio-dynamically. Now the two diseases referred to have practically disappeared."

Farm E. "Before conversion there was a great deal of sickness in the stable of this farm which—ironically—had belonged for three years to a member of an Association for the *Combating of Breeding Diseases.* It was also difficult to get the cows in calf. Up to the time of conversion there had been scarcely any improvement in the situation; after the conversion and treatment with Weleda remedies an improvement soon began. Since that time the cows have been easily bred."

Farm F. "... Until in 1931 *contagious abortion* was prevalent. This drove us, in the course of the year, to weed out all 'bought' cattle. There are now only a few of their offspring left. Since beginning here with the bio-dynamic method of agriculture, we decided to reform our herd entirely with local 'red and white' cattle. Our own breeding, since the conversion, has been very satisfactory. We give credit to the change in method, for the fact that, since then, there have been no more cases of abortion. Since that time the fertilization and the carrying time of the cows have returned to normal. Breeding diseases are absent."

Farm G. "As is the case in other bio-dynamic farms, the first phenomenon was a luxuriant legume growth. Exceptionally good clover, and harvests of 3600 lbs. of broad beans and 2500 lbs. of vetch per acre, were gathered. Primarily there was improvement in the cow byre. Although the purchase of forcing feed was completely given up, the milk production, with only home-grown feed, was raised by about 2000 lbs. per cow, and the butter fat content by about 12 per cent. Our total figures on feeding showed that with the same amount of 'starch units' we obtained after conversion about one fifth more milk and fat than previously.

"These figures proved to us that there is a greater value in

PRACTICAL RESULTS OF THE BIO-DYNAMIC METHOD

bio-dynamically raised fodder plants than in those grown with the customary agricultural practices. We feed the very last leaf of the fodder turnips without any sign of diarrhoea. Our neighbours feed no turnip tops, or very little, because they have found there is danger of diarrhoea from their use. The general health of the herd has also improved most satisfactorily. In the past year we were able to raise 20 calves from 21 cows."

We now present a series of examples of milk production before and after the conversion. The figures, with two exceptions, are taken from the reports of professional milk-production testing organizations.

Milk production per cow

	Year	Lbs. of milk	Percentage of butter fat	Remarks
1.	1929–30	9116		Strong epidemic abortion delayed calving.
	1930–31	8197		
	1931–32	6420		
	1932–33	5849		Two-thirds of the fields converted to the bio-dynamic method.
	1933–34	7460		
	1934–35	10498		
2.	1914–24	4409–6614		Before the conversion, production never exceeded 6614 lbs.
	1924–25	6581.06		After the conversion to the bio-dynamic method of agriculture.
	1925–26	6938.56		
	1926–27	8246.28		
	1927–28	8207.72		
	1928–29	7411.09		
	1929–30	7113.36		
	1930–31	7419.80		
	1931–32	7811.44		
	1932–33[1]	7113.36		
	1933–34	6220.16		Drought.

[1] In recent years the herd has been increased with its own heifers. The lowering of the average age of the herd, and the greater proportion of young cattle, with the drought, helped to lower the annual average. It is of course familiar that cows in their first lactation periods do not reach their full production.

Milk production per cow (continued)

	Year	Lbs. of milk	Percentage of butter fat	Remarks
	1934–35	6429.60		
	1935–36	6478.94	3.40	
3.	1929–30	9341	3.60	
	1930–31	9643	3.27	
	1931–32	9156	3.43	
	1932–33	9480	3.28	Conversion to bio-dynamic method.
	1933–34	10701	3.36	
	1934–35	9899	3.16	Pasture dried up.
	1935–36	10827	3.23	
4.	1931–32	6354		In process of conversion.
	1932–33	6607		
	1933–34	7418		Conversion completed.
	1934–35	6867	3.59	Drought.
5.	1932	6173	3.28	Conversion.
	1933	6393	3.24	
	1934	7496	3.17	
	1935	7055	3.11	Drought; in other farms in the local production-testing association, the average production frequently dropped some 1000 lbs.
6.	1930–31	6265	3.92	Conversion; there were still important changes taking place 1931–33, in the number of animals.
	1931–32	5747	3.95	
	1932–33	5456	4.11	
	1933–34	6196	3.71	
	1934–35	7789	3.84	
	1935–36	8527	3.73	

PRACTICAL RESULTS OF THE BIO-DYNAMIC METHOD
Milk production per cow (continued)

	Year	Lbs. of milk	Lbs. of butter fat	Remarks
7.	1932	8541	274.25	
	1933	8269	270.28	Conversion.
	1934	9028	293.87	
	1935	9901	313.71	
8.	1928–29	4696	179.71	
	1929–30	5567	210.84	Conversion.
	1930–31	6213	228.61	
	1931–32	5811	218.43	
	1932–33	6620	251.03	
	1933	6662	248.61	
	1934	5683	204.12	Drought.
	1935[1]	5485	203.19	Drought.

	Year	Lbs. of milk	Percentage of butter fat	Remarks
9.	1928–29	8016	3.77	Conversion begun.
	1929–30	8091	3.73	
	1930–31	8882	3.67	
	1931–32	9156	3.81	
	1933	8792	3.81	
	1934	9321	3.62	
10.	1929–30	6257	3.72	
	1930–31	5710	4.06	
	1931–32	6530	4.20	Conversion completed.
	1932–33	7639	4.21	
	1933–34	6629	4.37	
	1934–35	7513	4.08	Drought.
	1935–36	7756	4.18	
11.	1928–29 to 1933–34	7344	3.09	Before conversion.
	1934–35	8327	3.11	After conversion.

[1] The number of cattle was increased by about 20 per cent, hence the total milk production mounted.

SOIL FERTILITY, RENEWAL AND PRESERVATION
Milk production per cow (continued)

Year	Lbs. of milk	Percentage of butter fat	Remarks
1935–36	9658	3.21	Before change, used an average of 980 lbs. linseed-oil cake per cow. After, only home-grown grain and hay.

Note: On this farm the effect of the new fodder quality was recorded in terms of figures. Before the conversion, 100 parts of starch units was transformed into 169 parts of milk. After the conversion 100 parts of starch units was transformed into 215 parts of milk.

All the farms whose records are quoted here are in Central Europe. There the bio-dynamic method of agriculture has been in use for the longest time, and we are better able to make a comprehensive survey. The farms referred to are located in the most varied conditions of soil and climate, and include places on the plains as well as others in hilly country. The averages of milk production given are those of the complete herds. The annual average of sixteen tested bio-dynamic farms in Germany was 7366 lbs., while at the same time the annual average for all Germany (all farms) was 5113 lbs., and that of all the tested farms (with 10 per cent of the total head of cattle in the country) was 7275 lbs. The data gathered in Germany apply correspondingly to the conditions of other countries in which there are bio-dynamic farms in operation.[1]

We have purposely chosen the milk-production figures because they give a decisive indication of the state of health of the cattle. This state of health, together with the manure produced by the cattle, represents the backbone of a biological agriculture. If the herd is in good condition, then the fertilizing can be brought into good condition and the fertility of the soil maintained.

Worthy of special mention is an estate farm situated east of Berlin.[1] It is located in the least advantageous conditions of

[1] In revising this edition the author would have liked to bring the data on milk production up to date. Owing to the war and other circumstances this was unfortunately not possible. However, the author's experiences in Holland before the war and since then in the United States confirm the upward trend

PRACTICAL RESULTS OF THE BIO-DYNAMIC METHOD

climate and soil. On a purely sandy soil in hilly country, surrounded by poor pine woods, this farm had rain as follows (given in millimetres):

1932	418	1935	436
1933	348	1936	330 Jan.–Oct.
1934	374		

Dew and mist and sub-surface water are equally scarce. This farm has 150 acres of cultivated land and 37 acres of meadows. The soil is 70 per cent pure sand. The bio-dynamic method was introduced along with the cultivation of mixed crops.

I. 1. Potatoes plus 11 tons of prepared stable manure per acre.
    2. Oats plus lupin or oats plus vetch plus hop clover (or summer rye plus summer barley plus sweet clover).
    3. Rye plus serradilla.

II. 1. Lupin.
    2. Rye plus 7 tons of prepared manure per acre; followed by lupin.
    3. Oats.

III. 1. Serradilla or lupin grown for seed.
    2. Rye plus 7 tons of prepared manure; followed by lupin.
    3. Oats or barley.

IV. 1. Turnips plus 14 tons of prepared stable manure.
    2. Summer barley plus clover.
    3. Clover.
    4. Rye or spelt (Triticum spelta) followed by lupin.

V. 1. Oats.

in milk production after the conversion. The figures from Holland are lost, the recent statistics on American farms are not yet complete. So far, however, it can be said that an increase in production of 30 per cent within two to three years with the same herd through conversion is not unusual. Holstein dairy cows have improved in the author's herd from 7,000 lbs. per lactation to 10,000 lbs. per lactation. Improvements have even been achieved with older cows which were already beyond the prime of their life. If for economical reasons one has to force production with additional grain feeding, one will observe that the 'basic' production on grass, clover, alfalfa hay and on improved pasture will rise from 5,500 lbs. per year to 7,000 or 8,000 lbs. and that the additional increase can be effected with grain feeding. It has been observed that a ratio of additional grain (18 per cent) or 1 lb grain per 4 lbs. of milk can be reduced to a ratio of 1 to 7 on improved pasture without any drop in production.

Dr. N. Rehmer, "Die Rindviehhaltung im Mittelpunkt eines Betriebsorganismus", *Demeter*, 1936, No. 10.

SOIL FERTILITY, RENEWAL AND PRESERVATION

The mixed crops, with a generous use of legumes, markedly increased the soil fertility, so that after a few years even oats and clover gave sure yields. Before the conversion the succession was: potatoes, rye, summer grain, and much mineral fertilizer. By the use of mixed crops a better ground cover was effected, with such beneficial results as an increase of moisture reserve and the shade it gave the soil. The increase of humus substances in the soil also helped its retention of moisture. In addition, this regime provided the fodder necessary for the manure supply. The capacity of the soil for growing clover was increased, which showed itself, after a few years, by the fact that lucerne did well there. By 1937 they kept 2½ acres planted with lucerne and 6¼ acres in clover. The potato acreage was lessened. Because of the cultivation of mixed feeds, it has been possible there to give green fodder in the stable from May to July. Clover was cut twice, lucerne up to three times; it was pastured from August on. The feeding was divided as follows; May to July, green cut rye mixed with clover and lucerne; July to October, pasture on serradilla, clover, lucerne and on grass meadows, and supplementary feeding on sunflower seeds. November, sunflower seeds, mixed hay, turnip tops, straw. December to April, turnips, straw, hay. Regarding the soil conditions, it must be noted that the souring of the soil was retarded not by spreading lime on the ground, but by the careful application of the sum of all the bio-dynamic measures.

In the course of a number of years the herd, which had been up to 70 per cent sterile, was cured. After the conversion to bio-dynamic procedures, there were born:

1932	from	22	cows,	23	calves
1933	,,	18	,,	17	,,
1934	,,	17	,,	16	,,
1935	,,	23	,,	22	,,
1936	,,	23	,,	23	,,

On this farm, too, as we have found almost invariably in our experience, it was again proved that the purchase of registered cattle from noted breeding establishments does not pay. The development of strong types, suited to the local conditions, progressed slowly but surely.

We cannot give all the details concerning this farm, which is of such special interest because it was run under such unfavourable conditions in the beginning. We shall, therefore, only add certain other figures. One of the cows on the place had, in fifteen years,

PRACTICAL RESULTS OF THE BIO-DYNAMIC METHOD

twelve calves of which five had already grown up to be valuable cows. Her milk production, in her fifteenth year, was still more than 9920 lbs. Of her bull calves, three were used for breeding purposes.

Average milk production of the farm per cow per year:

Before the conversion	No. of tested cows	Milk in lbs.
1926–27	10	5401
1927–28	11	7143
1928–29	10	6063
After the conversion	No. of tested cows	Milk in lbs.
1932	20	7055
1933 ⎫	16	7297
1934 ⎬ drought years	13 (lower average age of herd)	7385
1935 ⎭	14	7716[1]
1936	20	7275[2]

We also cite the individual development of certain young animals again. (A markedly strong bone development was noted in the bull calves on this farm.)

Year	B	C(i)	D(a)	E	F
		Lbs. of milk per year			
1932	8157	6360	6360	3931	—
1933	9094	6526	6788	6111	5414
1934	11005	7571	8488	6614	6991
1935	11354	7668	9524	8602	7745

If one keeps in mind that the development of the farm took place on a very sandy soil, this speaks more emphatically than any other fact for the practical usability of the new method.

As an example in the field of intensive gardening, we present the yields—which remained the same for six years—of an enterprise in Holland. In a greenhouse of about 1000 square yards, planted every year with tomatoes, 3000 plants produced from eight to nine tons (cf. page 102, Chapter X). In this connection it is to be noted that this yield came evenly for six successive years from the same piece of soil, and without plant diseases. A greenhouse with *cucumbers*—dimensions, 40 yards long, 3¾ yards wide—produced each year about 4400 green cucumbers. One of similar size produced about 2700 white cucumbers annually. A *grape hothouse*, 40 yards long and 8 yards wide, yielded, beginning with the third year, about a ton of grapes (Frankenthaler variety).

[1] 8157 for those tested in 1935.   [2] Entire herd.

SOIL FERTILITY, RENEWAL AND PRESERVATION

Let us consider a few more experiences on larger farms. Dr. B. von Heynitz writes[1] that, before converting to the bio-dynamic method, he paid out an average of fifty German Reichsmarks per hectare (2¼ acres) for chemical fertilizer, and about seventy to eighty Reichsmarks per hectare of land for additional concentrate feeding. He was not able to get yields corresponding to this outlay. So he started, experimentally, converting a part of his place and then converted the entire farm of 286 hectares (715 acres). He states: "I was particularly well able to evaluate the crops at this time, since in the first three years of the conversion a part of the fields was still chemically fertilized, while to an increasing extent the other section of the farm received bio-dynamic treatment. I was able to prove, in comparing yields, that the yields of the bio-dynamically fertilized crops were *not* quantitatively behind those produced with chemical fertilizer." An important phenomenon, observed everywhere, is that bio-dynamic grain shows less tendency to "lodge"[2] under moist conditions.

After the conversion the yields were:

	lbs. per acre	
Rye	2363	
Winter wheat	2985	
Summer wheat	3144	
Winter barley	3092	
Oats	2637	
Potatoes	21032	(18–22 per cent starch depending on the variety)
Sugar beet	20106	(18.8 per cent sugar)
Fodder turnips	77690	

These data concern a farm in hilly country with a heavy soil, in Saxony. The use of concentrated feeding is interesting; after the conversion, this cost 17 Reichsmarks per hectare as against 70 previously. This was found while maintaining the milk yield at a point of 6614 to 7716 lbs. per animal per year, with 3.25 per cent butter fat.

From another report, concerning an intensive grain-growing farm on a light clay and siliceous soil, we take the following

---

[1] Dr. B. von Heynitz, "Meine Erfahrungen mit der biol. dynam. Wirtschaftsweise und dem Absatz ihrer Erzeugnisse". *Demeter*, 1934, No. 2.

[2] "To lodge" = to remain down after being beaten down.

PRACTICAL RESULTS OF THE BIO-DYNAMIC METHOD

data.[1] We are dealing here with a definitely poor farm. Two earlier tenants went bankrupt, a third paid no rent, a fourth was barely able to keep going. The conversion to the bio-dynamic method proceeded slowly in the course of five years:

Yields

Year	Total acreage	Total acreage bio-dyn. treated	Yield in lbs. per acre
			Wheat
1924–25	165	—	1834
1925–26	107	—	1640
1926–27	210	—	1852
1927–28	107	—	1799
1928–29	115	—	2840
1929–30	130	—	2183
1930–31	215	27	1481
1931–32	157	110	1949
1932–33	140	140	2319
			Rye
1924–25	157	—	2266
1925–26	87	—	1675
1926–27	97	—	1808
1927–28	200	—	2169
1928–29	107	—	2390
1929–30	112	6.3	2319
1930–31	72	20	1684
1931–32	95	32.5	2390
1932–33	152.5	152.5	2205
			Winter Barley
1924–25	—	—	—
1925–26	20.3	—	1182
1926–27	13.3	—	2884
1927–28	17.5	—	2690
1928–29	40	—	2346
1929–30	55	—	2857
1930–31	70	—	2610
1931–2	67.5	6.2	2593
1932–33	67.5	67.5	2699

[1] Dr. A. Vogelsang, "Der Betrieb Rittergut Böhla", *Demeter*, 1934, No. 2.

SOIL FERTILITY, RENEWAL AND PRESERVATION

Yields

Year	Total acreage	Total acreage bio-dyn. treated	Yield in lbs. per acre
			Oats
1924–25	95	—	1808
1925–26	220	—	2801
1926–27	120	—	2222
1927–28	150	—	1870
1928–29	132.5	—	2328
1929–30	100	—	2028
1930–31	82.5	—	2399
1931–32	57.5	—	1870
1932–33	62.5	57.5	1808
			Potatoes
1924–25	95	—	10847
1925–26	145	—	12875
1926–27	122.5	—	13757
1927–28	127.5	—	8871
1928–29	122.5	6.2	12346
1929–30	142.5	32.5	13510
1930–31	140	140	13424
1931–32	125	125	15450
1932–33	122.5	122.5	15609
			Sugar Beet
1928–29	6.2	—	23300
1929–30	10	—	32804
1930–31	15	—	23810
1931–32	15	15	20635
1932–33	15	15	21870

Taking into consideration weather differences of the individual years, the fluctuation and occasional decline in yields are not of major significance, since soil tests, which were carried on simultaneously with the keeping of the crop records, showed no decline in the nutritive materials in the soil.

A small farm of 57 acres on medium and light soil, in a region poor in rain, reports:

Average yield in pounds per acre:

	Before conversion	After conversion
Wheat	1360	1640

PRACTICAL RESULTS OF THE BIO-DYNAMIC METHOD

	Before conversion	After conversion
Rye	960	1160
Oats	1560	1720
Barley	1520	2000
Maize	2360	3000
Fodder beans	1920	1840

It is a general experience that small farms can be converted more quickly and simply than the very large ones. The former are generally fertilized intensively—that is, they have more cattle per acre, as was the case in the one just cited. Here the careful, bio-dynamic handling of the manure has especially beneficial effects, which explains the relatively high yields.

These experiences, gathered under pre-war continental European conditions, are directly applicable also for England and America. It is the author's firm conviction, on the basis of personal studies on various visits to England and of practice for the last five years in America, that the standard of agricultural work in these countries could, by means of careful work, be greatly raised. This would, of course, require the co-operation of a great many farmers. The proper procedure in this connection would be first to set up, in various locations in both countries, model and demonstration farms. These could then be visited by all farmers. On these farms the adaptability of the methods to the most varied conditions of climate and soil could be demonstrated. It is a familiar truism that the farmer learns more through practical demonstrations than from any amount of writing, because he is *more convinced* by what he sees than by what he reads. So this book aims to constitute a brief instruction in how to make possible this "being able to see the thing in practice". It is no textbook or directive on farming. Rather it presents the long-neglected points of view which will, when put into practice, restore ordinary farming to a healthy state which in turn will help to foster a permanently sound agriculture and a healthier human society.

CHAPTER FIFTEEN

Man's Responsibility

European agriculture is on the verge of surrendering its traditional methods and of abandoning the conscious carrying on and the working and directing of its farms. Yet the European still has "solid ground" under his feet which permits of an almost biologically balanced agriculture and in this lies his chance of a hopeful future. It is true that in Europe the biological possibilities of the land have already been overstepped, but this has happened in a fashion which still permits of a turning back and a renewal.

Europe as a middle region between Orient and Occident represents a sort of centre of gravity in the previously described tendencies of the Far East and the Far West. Here life conditions exist out of which a new building up of a natural basis for food supply can be developed. Here, in a healthy way, methods might be devised and biological laws recognized which would be beneficial to the people of Europe, indeed, to the whole West, as well as to the diseased condition of the earth's organism in the Orient. Here outer nature extends, so to speak, a helping hand.

However—and here we come to a point which will perhaps appear surprising to many—the solution does not depend on nature, but on the human being. The solution of the agricultural crisis of the present time is a human spiritual problem. It consists in man's extending his knowledge of nature's being, of life's laws, and in the *creation of a method of thought founded on the principle of an Organic Whole*.

If once the foundations of traditional culture are lost—a process which is becoming more evident in all spheres of life as the fundamental trend of the twentieth century—then no turning hither or thither, no probing, no clever thinking things out, discovering and applying, will help. All study and all that we may do remain mere patchwork, *as long as the one great task, the creating of a new culture, is ignored, is not recognized.*

MAN'S RESPONSIBILITY

If the new work we do is not be the mere appending of the thousand and second case to the thousand and first, but rather the means of gathering and presenting a body of knowledge imbued with life and the capacity to develop, then a way can be found out of our difficulties. If we provide some one particular foodstuff, grow a specific plant variety of grain, indeed, if we do something which in its details fills our day as an agriculturist, it will all remain patchwork if these efforts are not based on a fundamentally different attitude, on our part, to the problems of life and growth —*an attitude which allows us to perceive life and growth as an organic whole over the entire earth.*

Someone will object that this is all very well, but has nothing whatever to do with agriculture as such. The answer is that the one-sidedness of the prevailing points of view about life, nature, and the universe has been one of the chief reasons for the collapse of our culture, as the events of our times themselves fully demonstrate. Another will say: "Very well, we acknowledge the broad, constructive advance implied in your first, incomplete presentation of this point of view; let us go on and rearrange everything on that basis." The best reply to such a person is, that the finest lecture on soil treatment is no help at all if, when the attempt is made to apply what is given, it is found that the farmhand, working in the field, does not know how to plough. Metaphorically speaking, we have to-day too many "lectures" and too few "farmhands". With these thoughts in mind, our book can be completed with the following summary.

Seeds need time for developing their qualities. They are planted repeatedly according to the rhythm of crop rotation. We know in agriculture that when the same crop is planted in the same plot of ground after three, five, seven, or more years, i.e. in long periods, only then does the value of crop rotation become evident. We learn to observe and work in terms of long rhythms of development. The forester's rhythms are stretched out even further in time. The goal—even when clearly known and set up—can be attained only in the course of the rhythm of development. *This is the first basic truth:* A curing of the ills of agriculture cannot be attained between to-day and to-morrow nor in one, two, or four years. Since this matter is concerned with processes of growth, the rhythms and times of these must be noted and strictly followed. Whoever is working to a plan for bringing an individual farm back to health knows and takes this into account.

The *first step* on such a farm is to gather together and carefully

SOIL FERTILITY, RENEWAL AND PRESERVATION

nurture all organic fertilizer material which may be at hand. This alone gives the basis for a production of humus in the soil.

The *second step* is concerned with proper cultivation of the soil. If the humus already exists, it is only necessary that this be maintained. The correct, soil-conserving crop rotation is the important thing there. Although the first step can be completed in one year, the second requires a period of from four to eight years.

The *third step* is the improvement of the cattle, for these in turn furnish the "raw material" for the first step. What this means is that the repetition of the first step—the gathering of organic fertilizer—has more meaning later on because it takes place on a higher level of quality, for it is produced by the farm's own, improved cattle.

The *fourth step* is the biological reshaping of the farm's environment, its biological area as a whole. Presumably, in the meantime, we have gathered sufficient wisdom for this through observation and experience. Keeping step with the process of development on the individual farm and applying the same principle which can give us a living understanding of a whole area of country (after we have tried it out in miniature) means that we do not seek a solution for one or another individual difficulty by itself, but work so that the programme as a whole is evolved further and made healthy. Perhaps the most significant words in the resolution of Sir Merrick Burell were: "Only a carefully thought out, long-term agricultural policy"!—not starting by producing tractors and then having no one to run them, as was said of Russia.

Goethe and Rudolf Steiner are milestones on this road of development. Goethe for the first time enunciated the laws of the organic, the superior whole, and the method of cognition belonging to it, in a conscious form in harmony with exact thinking; Rudolf Steiner carried the Goethean idea further until at last it is in a form which the practical agriculturist, the farmer, can grasp. Elsewhere in this volume we have described the practical application of the bio-dynamic method of agriculture, a method which consists, first of all, in learning to understand the laws which govern the perception of the living, the functioning of the life process: that is, we have been "Goethean" in the development of our ideas.

The farmer who wishes to convert his farm in accordance with the new bio-dynamic points of view has first to work on himself, to learn to think along different lines. This is the greatest difficulty involved in introducing marked innovations; people would like very much to have an automatic recipe for getting their work done without having to help with their own inner activity. In the practical spheres of life, such as farming and gardening, or forestry, this is of course impossible. The human being himself is the strongest nature force, which guides and directs the beginning, the course and the end of the natural growth process. His capacity is the final decisive factor.

We have known of more than one farmer who began deepening his interest in what went on in the fermentation of the manure and compost and suddenly came upon the most significant discoveries concerning the formative forces in the sphere of life. A good example of this is to be seen in the way good bread is made. A ripe manure makes a crumbly soil. The grain ripens on it properly. After mowing, it goes through an after-ripening process in the ear. If it has been threshed immediately after mowing, this ripening process cannot take place. Furthermore, the best quality is retained by storing the grain in the ear rather than in the form of threshed-out seeds: the beetles and mice do less harm. We find that late-threshed grain, which has ripened in the ear, gives better bread. The threshed grain "sweats"—it is still alive—it "ripens" further. When it has come to a state of rest it can be ground. Then it is in best condition for baking, though the flour still "works" on and needs a further three to five weeks to reach the peak of its baking value. Are there any who still remember these things? They were known once! Who considers these things now? How often are chemists and technicians called in to help, to get rid of the defects arising out of forgetfulness and ignorance!

The bio-dynamic farmer is himself trained to pay attention to the finest processes. But this presupposes that in him there is taking place a slow, inner change. As he learns to recognize and comprehend in a deeper sense the value and the potentialities for development of the life process by a "rotation" of the spiritual processes of perception, gradually out of the "soil-tiller mechanic" there develops again a real farmer. There develops an ethical feeling of responsibility toward that organism, "the living soil". Reverence for life as a whole develops—that is, an inner relationship to the calling of "tilling the earth" is unfolded. To-day the bio-dynamic farmer is inwardly upheld by conscious knowledge—formerly it was instinctive tradition—but outside of this way of working there is neither consciousness nor tradition. Herein lies the founding of a new "peasantry". This conscious knowledge is the only guide that can lead to a healing of the world-wide sickness of agriculture and society.

SOIL FERTILITY, RENEWAL AND PRESERVATION

The "farming industrialists" and "rent-chasers" get restless at this point. They have no idea, in any case, how to improve the conditions of agriculture. *But that this should take place by way of the inner spiritual attitude of the individual and his relationship toward his calling*—well, my dear Sir—that is asking altogether too much! But in the meantime farms are being cultivated which have already shown, on the basis of ten and more years' experience, that the new point of view and approach has already affected the farmer. He has achieved the stamping out of the "diseases" of his stable, and in dry periods his fields form a green island in the brown landscape, and, above all, he is learning to love his soil again.

An inner and an outer transformation in accordance with the laws of life was the first result of the Bio-Dynamic Method of Agriculture. Numerous farms stand witness to this—farms which have carried on this work correctly. But the decision regarding a revolution in his habits of work and thought is something which must be left to the decision of the individual farmer.

Were you to question an experienced bio-dynamic farmer to-day concerning the solution of the world agricultural problems, his answer would have to be: No short-term programme, but a plan looking ahead for two generations; slow turning over of the land from incorrect to correct practices; and, even before a beginning is made on the working of the ground, a schooling of the human beings involved is needed. Those who desired to work the soil would have to look about to find people with open-minded attitudes and skill and capacity for the conscious direction and shaping of the biological cycle. These people—trained down to the smallest practical details—could go out to any individual farmer, in order to live with him and exemplify, on the spot, the true "peasant" life in its hard, daily effort and toil. This would be the first step in a plan embracing perhaps two generations—adult education in its truest sense. If this problem is solved, then the bringing back of the soil to health is really already accomplished. Whoever understands this has the key in his hand. *Farmer, in your hands lies the future!*

Bibliography

Anthony, R. D.: *Soil Organic Matter as a Factor in the Fertility of Apple Orchards.* The Pennsylvania State College School of Agriculture and Experiment Station, State College, Pennsylvania. Bulletin No. 261, January 1931.
*Baker, C. A.: *The Labouring Earth.* Heath Cranton. London.
*Baker, O. F., Borsodi, Wilson, and Ralph, M. L.: *Agriculture in Modern Life.* Harper Brothers, New York and London, 1939.
*Balfour, Lady E. B.: *The Living Soil.* Evidence of the Importance to Human Health of Soil Vitality, with special reference to Post-War Planning. Faber & Faber, London, 1943.
Barlow, K. E.: *The Discipline of Peace,* Faber & Faber, London.
Bouyoucos, George J.: *Rate and Extent of Solubility of Minerals and Rocks under Different Treatments and Conditions.* Agricultural Experiment Station of the Michigan Agricultural College. Technical Bulletin No. 50, July 1921.
Chase, Stuart: *Rich Land, Poor Land.* Whittlesey House, New York, 1936.
Conserve Your Soil; a simple plan for erosion control, published by the Bank of New South Wales, Sydney, Australia.
Daniel, A., Laugham, and Wright H.: "The Effect of Wind Erosion and Cultivation on the Total Nitrogen and Organic Matter of Soils". *Journal of the American Society of Agriculture,* Vol. 128, No. 8, 1936.
Daniel, Harley A.: "These Soils". *Journal of the American Society of Agronomy,* Vol. 128, No. 7, 1936.
*Darwin, Charles: *Darwin on Humus and the Earthworm,* Faber & Faber, London, 1945.
Elliot, Robert H.: *The Clifton Park System of Farming,* Faber & Faber, London, 1943.
Fagan, F. N., Anthony, R. D., and Clarke, W. S., Jr.: *Twenty-five Years of Orchard Soil Fertility Experiments,* The Pennsylvania State College School of Agriculture and Experiment Station, State College, Pennsylvania. No. 294, August 1933.
*Faulkner, Edward H.: *Plowman's Folly.* University of Oklahoma Press.

BIBLIOGRAPHY

Fippin, Elmer O.: *The Soil: Its Use and Abuse.* The Cornell Reading-Courses (Lesson for the farm), Vol. I, No. 2, The Soil Series No. 1, 15th October 1913.

*Fippin, Elmer O.: *Nature, Effects, and Maintenance of Humus in the Soil.* The Cornell Reading-Courses (Lesson for the farm), Vol. III, No. 50, The Soil Series No. 3, 15th October 1913.

Fippin, Elmer O.: *Tilth and Tillage of the Soil.* The Cornell Reading Courses (Lesson for the farm), Vol. II, No. 42, The Soil Series No. 2, 15th June 1913.

*Gardiner, Rolf: *England Herself,* Faber & Faber, London, 1943.

*Graham, M.: *Soil and Sense,* Faber & Faber, London, 1941.

Gustafson, A. F.: *Organic Matter in the Soil.* Cornell Extension Bulletin No. 68, published by the New York State College of Agriculture at Cornell University, Ithaca, New York, October 1923.

*Howard, Sir Albert: *An Agricultural Testament,* Oxford University Press, London.

*Howard, Sir Albert: "The Manufacture of Humus by the Indore Process". *Journal of the Royal Society of Arts,* Vol. LXXXI, No. 4331, London, 22nd November 1935.

*Howard, Sir Albert: *Farming and Gardening for Health or Disease,* Faber & Faber, London, 1945.

*Howard, Albert and Yeshwant, D. Wad: *The Waste Products of Agriculture; Their Utilization as Humus.* Oxford University Press, London, 1931.

*Jacks, G. V. and Whyte, R. O., *Erosion and Soil Conservation.* Imperial Bureau of Soil Science, No. 36, 1938.

*Jacks, G. V. and Whyte, R. O.: *Rape of the Earth,* Faber & Faber, London, 1939.

Jenkins, S. H., Ph.D.: *Organic Manures.* Imperial Bureau of Soil Science, Technical Communication No. 33, Harpenden, England, 1935.

*Jenny, Hans: *A Study on the Influence of Climate upon the Nitrogen and Organic Matter Content of the Soil.* University of Missouri, College of Agriculture, and Agricultural Experiment Station, Research Bulletin 152, November 1930.

*Jenny, Hans: *Soil Fertility Losses under Missouri Conditions.* University of Missouri, College of Agriculture, and Agricultural Experiment Station, Research Bulletin 324, May 1933.

*Kallet, A. and Schlink, F. J.: *100 Million Guinea Pigs.* New York, 1933.

BIBLIOGRAPHY

King, F. H., D. Sc.: *Farmers of Forty Centuries of Permanent Agriculture in China, Korea and Japan*. Jonathan Cape, London, 1926.
*Ligutti, Luigi, G. and Rawe, John C.: *Rural Roads to Security*. The Bruce Publishing Co., Milwaukee, 1940.
Lord, Russel: *To Hold This Soil*. Miscellaneous Publication No. 321, U.S. Department of Agriculture, 1938.
*Lymington, Viscount: *Famine in England*. H. F. & G. Witherby, London.
MacLeish, Archibald: *Land of the Free*. Harcourt, Brace & Co., New York, 1938.
McArthur and Coke, I.: *Types of Farming in Canada*. No. 653, Department of Agriculture, Canada, 1939.
*McWilliams, C.: *Ill Fares the Land*, Faber & Faber, London, 1945.
*Northbourne, Lord: *Look to the Land*. Dent, London.
News Sheet of the Bio-Dynamic Method of Agriculture, No. 3, November 1936. Enquiry Office, King's Langley Priory, Herts.
*Pfeiffer, Ehrenfried: *Formative Forces in Crystallization*. Rudolph Steiner Publishing Co., 54 Bloomsbury Street, London; and Anthroposophic Press, New York, 1936.
Pfeiffer, Ehrenfried: *Sensitive Crystallization Processes:* a Demonstration of Formative Forces in the Blood. Verlag Emil Weises Buchhandlung, Dresden, 1936.
*Pfeiffer, Ehrenfried: *New Methods in Agriculture and Their Effects on Foodstuffs. The Biological-Dynamic Method of Rudolf Steiner*. Rudolf Steiner Publishing Co., 54 Bloomsbury Street, London, 1934.
Piper, C. V. and Pieters, A. J.: *Green Manuring*. U.S. Department of Agriculture, Farmers' Bulletin No. 1250. Washington, April 1925.
*Price, Weston A.: *Nutrition and Physical Degeneration*. Paul B. Hoeber, New York and London.
Rayner, M. C.: *Mycorrhiza*, Cambridge University Press, 1927.
Rodale, J. I.: *Pay Dirt*. The Devin Adair Co., New York, 1945.
*Sears, Paul B.: *Deserts on the March*. Simon Schuster, New York, 1937.
Secrett, F. A., F.L.S.: "Discussion: Proceedings of the Society". *The Journal of the Royal Society of Arts*, Vol. LXXXIV, No. 4334, 13th December 1935.
Slipher, J. A.: *The Management of Manure in Barn and Field*, Bulletin 131, Agricultural College Extension Service, The Ohio State University, Columbus, Ohio. May & June 1914.
Soil; The Nation's Basic Heritage. U.S.A. Department of Agriculture.
Stapledon, R. G.: *The Land, Now and To-morrow*. Faber & Faber, London, 1935.

BIBLIOGRAPHY

*Sykes, Friend: *Humus and the Farmer*, Faber & Faber, London, 1946.

Waite Agricultural Research Institute, Report of, Glen Osmons, South Australia. University of Adelaide, 1937.

Webb, Walter Prescott: *The Great Plains.* Ginn & Co., New York and London, 1931.

Weir, Walter W.: *Soil Erosion in California: Its Prevention and Control.* University of California, College of Agriculture and Agricultural Experiment Station. Berkeley, California. Bulletin 538, August 1932

Wiancko, A. T., Walker, G. P., Mulvey, R. R.: *Legumes in Soil Improvement.* Purdue University Agricultural Experiment Station, Lafayette, Indiana. Bulletin No. 324, July 1928.

*Wrench, G. T.: *The Wheel of Health*, Daniel, 1938.

*Wrench, G. T.: *Reconstruction by Way of the Soil*, Faber & Faber, 1946.

Abderhalden, Prof. Dr. Med. et phil. h. c. Emil: "Bisher unbe-. kannte Nahrungsstoffe und ihre Bedeutung für die Ernährung". Halle A. S. *Zeitschrift für Schweinezucht*, Heft 17, 1922.

Abderhalden, Prof. Dr. Med. et phil. h. c. Emil: "Nahrungsstoffe mit besonderen Wirkungen unter besonderer Berücksichtigung bisher noch unbekannter Nahrungsstoffe für die Volksernährung". Published in *Die Volksernährung, Veröffentlichungen aus dem Tätigkeitsbereiche des Reichsministeriums für Ernährung und Landwirtschaft.*

*Ammon, Walter. Paul Haupt, Berne, 1937. *Das Plenterprinzip in der Schweizerischen Forstwirtschaft.* (Mixed forest principle in Switzerland.)

*Bartsch, Dr. Erhard: *Die Not der Landwirtschaft, ihre Ursachen und ihre Ueberwindung*, Emil Weises, Dresden.

Blank: *Handbuch* der Bodenkunde, Bd. I bis VIII, 1929.

Demeter, Monatsschrift für biologisch-dynamische Wirtschaftsweise. Bad Saarow, 1928–37.

*Dreidax, Franz: "Der Regenwurm", aus Koschützki, *Rationelle Landwirtschaft*, Berlin.

*Dreidax, Ludwig. *Untersuchungen ueber die Bedeutung der Regenwuermer fuer den Pflanzenbau.* Julius Springer, Berlin, 1931. (The importance of earthworms for plant growth.)

*Kapff, Prof. Dr. S. v.: "Von der künstlichen chemischen Düngung zur natürlich-biologischen Wirtschaftsweise", *Demeter*, 10 Jahrgang, No. 8.

BIBLIOGRAPHY

Krafft: *Lehrbuch der Landwirtschaft.*
Liek, Dr. L.: "Der Einfluss, der Düngung auf die Zusammensetzung der Nahrungsmittel". *Leib und Leben,* München, August 1935.
*Linstow, O. v.: *Bodenanzeigende Pflanzen.* Preuss. Geol. Landesanstalt, Berlin, 1929. (Soil-indicating plants. There is nothing existing on the subject comparable to this unique book.)
*Niklewski, Prof. Dr. B.: "Ueber den Einfluss von Kolloidstoffen auf die Entwicklung einiger Kulturpflanzen". *Jahrbücher für wissenschaftliche Botanik,* 1933, Band LXXVIII, Heft 3. Verlag von Gebrüder Bornträger. Leipzig.
*Niklewski, Prof. Dr. B.: *Zur Biologie der Stallmistkonservierung. Centralblatt für Bakteriologie, Parasitenkunde und Infektionskrankheiten.* Band 75. Verlag Gustav Fischer, Jena, 1928.
*Niklewski, Prof. Dr. B.: *Der Einfluss der Kompostdüngung und Behäufelung der Pflanzen auf Ernteproduktion (Streszczenie),* Warschau 1929. Summary in English.
*Pfeiffer, Ehrenfried: "Wind, Luft und Staub als bodenbildende Faktoren". *Kalender* 1934-5, mathem.—astron.—Sektion am Goetheanum, Dornach, Schweiz.
*Pfeiffer, Ehrenfried: "Een nieuwe methode om betere landbouwproducten te krijgen". *Natur en Techniek,* No. 10, October 1934.
Rost, Prof. Dr.: "Ueber Schwanz- und Fussgangraen bei Ratten". *Münchener Medizinische Wochenschrift,* 76. Jahrgang, Nr. 22. 31. Mai 1929.
*Schwarz, M. K.: *Ein Weg zum praktischen Siedeln.* Pflugscharverlag Klein, Vater und Sohn, Düsseldorf, 1933.
*Stoeckli, A.: "Die besonderen Wirkungen der sog. Humusduenger". *Schweiz. Landw. Monatshefte,* XII, No. 12, 1934. (The special effects of humus fertilizers.)
*Süssenguth, Dr. A., München: "Aus den Grenzgebieten der Medizin; Pflanzenernährung und Volksgesundheit". *Deutsche Medizinische Wochenschrift,* Verlag Georg Thieme. 59. Jahrgang, Nr. 51, Leipzig, 22nd December 1933.
Wölfer: *Grundzätze und Ziele neuzeitlicher Landwirtschaft,* Verlag Paul Parey, Berlin.

This list is not complete.
Books quoted in the text are not listed here.
Publications bearing especially on our subject are marked with an asterisk.

Index

Abies, 125
Abortion, 48, 72; contagious, 48, 73, 170
Acacia cebil, 117
Acer campestris, 125
Agriculture, traditional, 16; economic, 20
Albumen, 100
Aleurites, 125
Alfalfa, 68, 77, 99, 129, 176
Algae, 117
Alpine meadows, 36
Aluminates, 44
Aluminium phosphate, 116
Ammonia, 154
Ammonium nitrate, 115, 147; sulphate, 41, 50, 115; sulpho-nitrate, 157
Amygdalus, 125
Aphides, 103
Apples, 108
Arsenic, 41, 59, 102
Asparagus, 68
Atriplex hortensis, 127
Auximones (auxin), 108, 112, 142

Ball moss, 121
Banana, 105
Bang's disease, 48, 169
Barley, 117, 129, 163, 167, 175, 178, 181
Beans, 68, 70, 81, 98, 99, 181; broad, 70, 102-3, 117, 167; dwarf, 99, 103, 152; kidney, 132; runner, 97; soya, 117, 136
Beech, 90
Beet, 69, 98, 99, 166-7
Beetles, 105, 185; asparagus, 103; flea, 103; Mexican bean, 103
Berry bushes, 97, 106
Beta vulgaris, 125
Betula alba, 125
Bio-dynamic preparations, 70, 71, 76, 82, 84, 101, 102-3, 105, 106, 111, 112, 130 *seqq.*, 133, 136-7, 139, 142-3, 145 *seqq.*
Birds, 93, 103-4
Bollworm, 104
Bone meal, 58
Boric acid, 141
Boron, 117, 126, 141
Bracken, 58

Bran, 78
Brassica rapa, 125; Napus, 125
Broom, 94; sedge, 153; common, 124
Brussels sprouts, 98
Buckwheat, 124, 128
Butter fat, 74, 170
Buttercup, 153

Cabbage, 77, 98, 101, 103; butterfly, 102; lettuce, 98; stalks, 61
Cacti, 127
Caesium, 126
Calcium, 21, 114, 123-5, 126, 146-7; nitrate, 115, 116, 147, 154
Calf, paralysis, 170; pneumonia, 170
Cambium, 105
Camomile, 52, 99; German, 127
Canada, 20
Canals, 96
Canary Islands, 34
Cannabis, 125
Capillarity, 66-7, 120
Capsid bug, 104
Carbolic acid, 105
Carbon, 39
Carbonic acid, 31, 36, 113, 114
Carrot, 98, 99
Castanea, 125
Catalysis, 108
Catch crops, 69-71, 102, 103
Cattle, 30, 61, 64, 72, 153, 169, 174, 184
Cauliflower, 49, 98, 99, 100, 102
Celery, 98, 99, 102
Cellulose, 32
Central Europe, 22, 32, 45, 56, 61, 174
Chalky marl, 39
Chenopodiaceae, 117
Charlock, 129
Chervil, 98
Chickens, 81, 90, 149, 150, 152
Chilean nitrate, 115
China, 22, 23, 34-5, 97
Chlorine, 36, 120
Chlorophyll, 116, 119
Chrysanthemum segetum, 126
Clay, 21, 51, 58, 105

INDEX

Clover, 54, 69, 70, 77, 80, 81, 115, 116, 141, 170, 175; alsike, 79, 80; hop, 167, 175; red, 77, 79, 80, 129; sweet, 158, 175; white, 79, 80, 81
Clubroot, 103
Cochlearia Armoracia, 127
Cockchafer, 81, 104
Cocksfoot, 79, 153
Cocos, 125
Compost, 22–3, 55–60, 61, 71–2, 75, 78, 82, 84, 85, 91–9, 104, 105, 112, 113, 125, 127, 130–2, 133, 144, 152, 185; clay, 81; herb, 103; tomato, 101, 102; rotted-leaf, 102; straw, 103; half-rotted, 100
Copper, 41, 102, 105, 117, 123, 126; sulphate, 75, 97, 117
Corn, sweet, 97; billbug, 17
Cornflower, 128, 129
Cows, *see* Cattle; dairy, 61, 73–4
Couch grass, 129
Crested dogstail, 79
Crops, hoed, 81–2, 98, 127–8; manured, 98; mixed, 97–8, 176
Crop rotation, 72–4, 81–3, 84, 165, 166
Crowfoot, 111, 115
Cucumber, 57, 98, 99–100, 102, 177
Cultures, mixed, 22, 97–8, 127, 175–6
Cynosurus cristatus, 79
Cystisus scoparius, 124

Daisies, 124
Daktylis glomerata, 79
Dandelion, 52, 105, 111, 126, 129, 137
Datura stramonium, 126
Deforestation, 23
Dew, 33
Digitalis purpurea, 126
Ditch cleanings, 58
Ditches, 96
Drainage, 33, 41, 55, 58
Dust storms, 20, 35

Earthworm, 21, 38, 41, 49, 52, 58, 86, 132
Eggs, 150, 151
Elm, Dutch, disease, 33
Endive, 98
Enzymes, 108
Equisetum, 127; arvense, 105, 127
Erica carnea, 125
Erosion, 19–20, 86, 93
Espalier, 97
Esparcet, 128
Evergreen, 90; needles, 89

Fagopyrum, 125
Fagus silvatica, 124
Fennel, 99
Fermentation, 51, 56–8, 61, 67, 97, 108, 139, 151
Fern leafage, 90
Ferrous salts, 89; oxide, 89, 122

Fertilizer, chemical, 41–2, 43, 148, 162, 176
Fescue, 153
Festuca duriuscula, 117; pratensis, 79; rubra, 79
Flour, 185
Fodder, 71, 77, 105, 169, 176; legume, 72, 77
Forestry, 82, 89 *seqq.*, 185
Foxglove, 126
Frog, 108
Frost, 42, 70, 82, 85
Fungus, 112; disease, 100; pests, 101

Gangrene, 155
Garden cress, 98
Gardening, 96
Geotropism, 142–3
Germany, 32, 36, 174
Gluten, 140
Gourds, 57; seeds of, 117
Gossypium, 125
Grain, 69, 81, 84, 154, 158, 166–7, 175–6, 178, 185; summer, 64, 68, 77, 81
Grapes, 106, 177
Grape vine, 117
Grass, 54, 77, 78–81, 140, 163; English rye, 79; Italian, 79
Greenhouse, 101, 102, 114, 132, 146, 177
Greenland, 31
Growth hormones, 108
Gypsum, 36

Harrowing, 42, 66, 67, 71, 76, 81, 111, 127
Hay, 54, 77, 117, 176
Hayfield, 78–81, 82, 115
Heart rot, beet, 141
Heather, 21
Hedges, 33–4, 58, 96
Hedge trimming, 58
Hellebore, white, 153
Hemp, 103, 128
Henbane, 126
Herbs, 153; aromatic, 99; medicinal, 52, 129; root, 127–8
Herniaria glabra, 125
Hoeing, 75
Holland, 44, 78, 81, 115, 177
Hormones, 108, 139, 151
Horn, 58
Horseradish, 127
Horsetail, 52
Hot beds, 54
Humus, 19–20, 21–3, 39–42, 45–54, 58–70, 83, 86, 87–93, 100–16, 131, 132, 147
Hydrogen, 141

Ilex Aquifolium, 125
Insect grubs, 38
Iodine, 36

INDEX

Iris germanica, 117
Iron, 89, 116–18, 121, 122, 126; chloride, 146; oxide, 36
Italy, 95

Juglans, 125

Kainit, 148
Kale, 98
Knotweed, 129
Kohlrabi, 98, 99
Kaolin, 111

Lamium, 128
Larix, 125
Larkspur, 129
Lead, 41, 59, 97, 102, 117
Legume, 20, 21–3, 64, 67–72, 77, 78, 79, 98–9, 105, 115, 117, 132, 167, 170, 176
Lentil, 117
Lespedeza, 81
Lettuce, 98, 104, 146
Lichen, 97
Lice, plant, 103; wood, 55
Lime, 21, 41, 88, 146, 162
Limestone, 21, 39, 149
Linseed, 142
Linum, 125
Lithium, 127
Loam, 51
Loess, 35
Lolium perenne, 79; italicum, 79
Lupin, 57, 68, 71, 131, 175
Lupinus albus, 131

Maggots, 81
Magnesia, 122
Magnesium, 116, 124–5; sulphate, 126, 146
Maize, 77, 98, 99, 117, 147, 149, 181
Manganese, 117, 123, 126
Malt, sprouted, 169
Mangold, 78
Manure, 48, 51–4, 55–60, 61, 63–6, 68, 71–2, 73, 75–6, 78, 81–3, 89, 90, 98–100, 104–6, 111, 112, 130, 138, 139, 144, 151, 154, 184; chicken, 63, 102; cow, 51, 61, 105, 151; goat, 103; green, 70–1; heap, 49–55, 56–8, 59, 65; horse, 48, 51–2, 61; liquid, 50, 55, 56, 60, 62, 81, 100, 115; pig, 63, 102, 103; pigeon, 102; sheep, 151; stable, 49, 61, 64, 81, 100, 158, 167, 175; straw, 54, 55; unrotted, 49–50, 75, 104; meadow, 15–19, 73, 78–9, 81, 82, 105, 111, 115
Medicago lupulina, 79
Meteorites, 36
Mice, 154, 185
Middle West, American, 34
Milk production, 171–4, 177
Miscarriage, 169
Mississippi Basin, 19

Mole cricket, 103
Moorland, 25, 32, 33
Morocco, 35
Moss, 97; Spanish, 120, 122; wood, 90
Mulch, 96, 103
Mustard, hedge, 127, 129; black, 117; wild, 129

Naples, 90
Nasturtium, 103
Needles, evergreen, 90
Nettle, 52; dead, 127, 128; stinging, 89, 127, 128, 137, 158
Nightshade, black, 127
Nile Delta, 45
Nitric acid, 36
Nitrogen, 39, 50, 56, 62, 64–8, 71, 85–6, 87, 94, 96, 100, 105, 106, 114–16, 127
Nitrophosphate, 41

Oak, 124, 125
Oats, 70, 116, 117, 128, 143, 175, 178, 181
Oil, ethereal, 129
Onion, 99, 127
Orache, 127
Orange, 117
Orchard, 104
Oxygen, 39

Palestine, 77
Panicum miliaceum, 162
Pansy, wild, 129
Paper, 32
Parsley, 99
Pasture, 64–6, 72, 78–82, 84, 127; weeds, 56
Pea, 68, 72, 81, 82, 97, 99, 117, 167; vines, 54
Peanut meal, 169
Peat, 39, 50; ash, 36
Peppermint, 103
Peptone, 163
Phaseolus vulgaris, 125, 132
Phleum pratense, 79
Phosphate, 85, 89
Phosphoric acid, 36, 39–41, 45, 114–16, 118, 121, 126, 127, 133, 138, 162, 168
Phosphorus, 86
Phototropism, 142
Physicians' reports, 161–2
Picea excelsa, 124
Pig, 81, 84, 90
Pigeon, 162
Pig weed, 127
Pinus, 125
Pitch, 91
Ploughing, 65–7, 81
Poa pratensis, 79
Poppies, 128, 129
Potash, 35, 56, 138

INDEX

Potassium, 41, 44, 45, 85–6, 114–16, 119, 123, 127, 133, 155; chloride, 146, 167, 168, 176; nitrate, 154, 162; phosphate, 146
Potatoes, 51, 58, 61, 63, 66, 69, 77, 82, 98, 99, 100, 104, 117–19, 127, 130, 167, 168, 175, 176, 178
Prunus, 125

Quartz, 144
Quercus, 125

Radish, 98, 99, 103, 152
Rain-water, 36, 42, 93
Randia dumetorum, 117
Ranunculus, 115, 153
Rape, 125, 129
Rats, 163
Reeds, 57
Resin, artificial, 32
Robinia pseudacacia, 89, 94
Rock crystal, 144–5
Rocky Mountains, 95
Roots, 130
Roughage, 54
Rosemary, 103
Rubus fruticosus, 125
Rumex Acetosella, 126
Russia, 35, 184
Rye, 68, 109, 117, 129, 166–7, 175, 176, 178, 181

Sage, 103
Sahara dust storm, 34
Salad, European corn, 98
Salsify, 98
Salts, 113
Sambucus nigra, 117
Sand, 58, 105, 123, 132; alluvial, 38; argillaceous, 36
Sarothamnus scoparius, 124
Sawdust, 54
Saxony, 178
Scarecrow, 104
Scrophularia nodosa, 125
Seaweed, 58, 126
Seed, 130, 153, 159, 162, 165, 180–1; bath, 128, 136; culture, 74
Self-maintenance, 141–2, 143
Serradilla, 68, 70, 175, 176
Shantung, 22
Sheep's sorrel, 126–7
Silica, 38, 121, 124
Silicates, 44
Silicic acid, 121, 122, 126, 127
Silk, artificial, 32
Snails, 104
Soda, 122
Sodium chloride, 44, 126

Soil, 48, 131, 140, 165; acidity, 19–20, 90, 116; bacteria, 21, 38, 39; clay, 166; cultivation, 184; fertility, 21–2, 25–7, 83, 135, 176; moisture, 33; orchard, 59; sandy, 174, 177, 178; vineyard, 41, 59; woodland, 87
Solanaceae, 117
Solanum nigrum, 127; tuberosum, 125
Soluble salts, 91
Sorrel, 153; sheep's, 126
Spinach, 31, 82, 98, 104
Spiraea ulmaria, 125
Spirogyra, 117
Spraying, 97, 103, 105–6, 131, 146
Starch, 81
Stellaria media, 125
Straw, 54, 58, 61, 77, 87
Strawberry, 106; weevil, 104
Strontium, 126
Sugar cane, 82; beet, 76, 117, 126, 166–7, 168, 178
Sulphur, 122, 127
Sulphuric acid, 36
Sunflower, 143, 145–6, 176
Swamp, 33

Tadpole, 108
Tanglefoot, 105
Tannic acid, 90
Taraxacum officinale, 137
Theobroma, 125
Thorn apple, 126
Thrombosis, 155–6
Thyroid gland, 108
Tillage, 48
Tillandsia usneoides, 122, 131
Titanium, 117, 126
Tobacco, 82, 123, 138–9
Tomato, 98, 99, 102–3, 117
Tradescantia, 105–6
Trees, 104, 105; alder, 96; apple, 117; beech, 90, 91; birch, 91, 112, 124; citrus, 82; elder, 94, 112, 113; fruit, 97, 103–6; hazel, 94, 96; locust, 94; locust yellow, 89, 126; oak, 90, 91, 124, 125; oak scrub, 94; pear, 106, 117; pine, 90, 117; poplar, 96; spruce, 91; wild cherry, 96

United States, 93
Urine, 133
Urtica dioica, 127, 137; urens, 127

Valerian, 52, 128, 144
Vanadium, 126
Vetch, 57, 70–2, 170, 175
Vicia faba, 117
Vitamin, 139
Volcanic ash, 34; eruptions, 34; rock, 38

INDEX

Water level, 92–3
Weeds, 57, 58, 75, 90, 94, 112, 124–7
Wheat, 20, 68, 131, 140, 143–4, 148–9, 154–5, 162–3, 166–7; winter, 63, 129
Windbreaks, 74, 96, 97
Wood ashes, 58
Woodlouse, 104
Woods, 89 *seqq.*, 93, 95

Wormwood, 103

Yarrow, 111, 127, 128
Yeast, 52

Zeolites, 44
Zinc, 123

Appendix A

Page 55, Line 13 from the bottom, after the word 'sponge'
I) In Middle Europe it is often difficult to keep enough moisture in the compost heap. For conditions in the UK, compost heaps are best domed instead of having a depression along the top. The covering will then allow the rain to flow off.

Page 59, Line 2, after the word 'used'
II) Half slaked or hydrated lime is alright for use. Unslaked lime will not be obtainable. Lime in the compost heap helps control fungal and other undesired growth, which the compost might otherwise support, and it speeds up breakdown.

Page 61, 2nd Paragraph, Line 12, after the word 'horses'
III) Since horses are no longer used, more cattle can be kept in their stead with the same results as far as manure is concerned. The lack of labour on farms unfortunately reduces the sources of compost. On the other hand, a lot of material is available commercially, such as municipal compost and manure from commercial animal enterprises and from riding stables, albeit one might have to be a little careful with these because of poisonous substances they may contain, for instance wood preservatives. Machines can be used to set up the heaps and soil can be shoved up onto a heap in the field, as this soil then goes back again on the same land.

Page 80, Line 2, after the word 'authority'
IV) Herbal leys and Clifton Park mixtures were then probably unknown to the author, or he would have mentioned them. In Middle Europe, grassfields bring forth an abundance of herbs.

APPENDIX A

Page 80, 3rd Paragraph, Line 9, after the word 'types'
V) This applies to continental conditions, as we get only one cut of hay a year as a rule. Still, the time of cutting is important.

Page 97, 2nd Paragraph, Line 3 from the bottom, after the word 'avoided'
VI) Those were the sprays then used. They were comparatively harmless. Today's "more efficient" sprays have, incidentally, not reduced pests and diseases.

Page 185, Chapter 2, after the word 'ignorance'
VII) Today, unfortunately, we have to put the word 'necessity' for 'forgetfulness and ignorance'. However, this necessity developed from it. Combined grain, it is true, stands longer in the ear before cutting than did stooked and ricked grain. The most careful drying after combining is, however, a poor substitute for the ripening in ricks. Sprouting in the ear has been vastly reduced by modern methods, nor do rats and mice get into the non-existing ricks. Nevertheless, the baking quality of wheat harvested in the old way is vastly superior to any we can find now. The difference in weather, for instance, in the various years were evened out in the ripening in the rick, to a certain degree.

BOOK LIST 1983

A selection of books for reference or further reading.

1) Rudolf Steiner — 'Agriculture'
2)* Maria Thun — 'Work on the Land and the Constellations'
3)* Maria Thun — 'Working with the Stars' — annual sowing calendar
4) Koepf, Petterson, Schaumann — 'Bio-Dynamic Agriculture'
5) G. Corrin — 'Handbook on Composting and the Bio-Dynamic Preparations'
6)* K. Castelliz — 'Life to the Land'
7) G. Grohmann — 'The Plant'
8)* W. Cloos — 'The Living Earth'
9) The Pfeiffer Gardening Book
10) John Soper — 'Bio-Dynamic Gardening'
11) E. Pfeiffer — 'Weeds and what they tell'
12) Philbrick & Gregg — 'Companion Plants'
13) Philbrick — 'The Bug Book'
14) M. Geuter — 'Herbs in Nutrition'
15) Agnes Fyfe — 'Moon and Plant'
16) Agnes Fyfe — 'The Signature of the Planet Mercury in Plants'
17) Theodore Schwenk — 'Sensitive Chaos'
18) J. Bockemuhl — 'In Partnership with Nature'

'Star and Furrow'   Published in Great Britain twice yearly by the Bio-Dynamic Agricultural Association, Woodman Lane, Clent, Stourbridge, West Midlands, DY9 9PX, which also supplies the above titles.

'Bio-Dynamics'   Published in the U.S.A. quarterly by the Bio-Dynamic Farming & Gardening Association, Richmond Townhouse Road, Wyoming, Rhode Island 02898.

* Also available direct from Lanthorn Press, Peredur, East Grinstead, Sussex.